Best Map Projections

Elena Novikova

Best Map Projections

Springer

Elena Novikova
Department of Geodesy
Kryvyi Rih National University
Kryvyi Rih, Ukraine

ISBN 978-3-031-78333-3 ISBN 978-3-031-78334-0 (eBook)
https://doi.org/10.1007/978-3-031-78334-0

This Springer imprint is published by the registered company Springer Nature Switzerland AG
The registered company address is: Gewerbestrasse 11, 6330 Cham, Switzerland

Dedicated to my teachers: Ekaterina Zaydova and Adolf Butkevich

Preface

The middle of the sixteenth century: a dark time of the Inquisition, the fight against dissent, the banning and destruction of a huge number of books, and a time of wars and epidemics. Yet, it was also during this time that Guillaume Postel[1] lived—a great scientist, humanist, philosopher, and a person of deep erudition. He was knowledgeable in astronomy, geography, and a significant number of languages, both modern (at that time) and ancient. Most importantly, he excelled in **cartography**. In contrast to the cruel and unjust world in which he lived, he constructed his ideal world, one without wars and contradictions, where all people would live happily and in complete harmony with each other and with the world around them. To depict this world on a map, he used the projection that now bears his name: the Postel projection. **Incredible!** Much later, mathematicians discovered that Guillaume Postel had used an **ideal** projection for his **ideal** world. Today, Postel's humanistic ideas are

[1] The portrait of Guillaume Postel is from Les vrais pourtraits et vies des hommes illustres grecz, latins et payens (1584) by André Thevet, tome 3, folio 588 recto. https://commons.wikimedia.org/wiki/File.jpg. This work is in the public domain in countries where the copyright term is the author's life plus 100 years or fewer and in the United States. For more information, visit Public Domain Mark 1.0 and the United States public domain tag.

represented on the UN emblem in the form of a world map built on his projection. Unfortunately, even now, humanity is far from the ideas proclaimed many centuries ago by the outstanding humanist cartographer Guillaume Postel. The unworthy part of mankind continues to destroy their own kind, as well as all living things around them. However, the best part of humanity strives for perfection. In mathematical cartography, this desire leads to the task of finding ideal and best projections. This book is dedicated to this task.

Competing Interests

I, the author of the manuscript titled "Best Map Projections", declare the following:

1. Financial Interests:
   I declare that I have no financial interests or relationships that could influence the research, results, or conclusions presented in this manuscript. No funding from third parties or organizations with a potential conflict of interest has been received.
2. Professional Interests:
   I declare that I have no professional interests that could be considered a conflict of interest related to this research. I am not associated with any commercial entities that would benefit from the results of this research.
3. Personal Interests:
   I declare that I have no personal relationships or interests that could influence the objectivity of the manuscript.
4. Other Competing Interests:
   I declare that there are no other competing interests that could be perceived as influencing the results or the interpretations of this research.

By submitting this statement, I confirm that all information regarding potential competing interests has been disclosed and that there are no conflicts that would influence the integrity of the manuscript.

Kryvyi Rih, Ukraine Elena Novikova

Introduction

The history of creating the best projections can broadly be seen as the history of cartography, since every cartographer, when creating a map, first intuitively and then consciously, tries to minimize its distortions. However, following the recommendation of Loyd Brown (1949), regarding the need to limit the straight and more or less narrow path in the presentation of history, we will describe here only those works that are directly related to the development of the theory of best projections.

The first step in creating the theory of best projections was made not in cartography, but in mathematics. In 1750, scientists Leonhard Euler and Joseph Louis Lagrange developed the theory of the calculus of variations (Caparrini 2014), which allowed another great mathematician and cartographer, Airy (1861), to find the second ideal projection that bears his name, the Airy projection. Unlike Guillaume Postel, who found the ideal projection intuitively (Snyder 1987, p. 191), G. B. Airy had previously formulated a criterion for choosing an ideal projection, which is now called the Airy criterion. He then used the theory of the calculus of variations to rigorously solve the extremal problem of finding an ideal projection using the chosen criterion.

Unfortunately, the Airy projection was more mathematically complicated than the azimuthal projections known at that time. This is probably why, completely unfairly, it is now practically not used.

In 1881, the fundamental work of the French scientist Nicolas-Auguste Tissot appeared in print, in which he laid the foundations for the theory of distortions of map projections. In his work, he proposed a technique for measuring distortions on maps, which made it possible to choose exactly the projection in which the distortions are minimal. Therefore, 1881 can be considered the beginning of the creation of the theory of the best projections as a science.

Due to the fact that this work appeared in print later than Airy's work, the definitions developed and used by N. A. Tissot and accepted in modern times were not applied by G. B. Airy, as Lapaine (2016) points out.

The second half of the nineteenth century was marked in mathematics by the active development of problems in the theory of functions. Mathematical cartography proved to be a successful means of confirming the theorems of functional analysis and demonstrating how a seemingly "pure" theory can be applied in practice. This is how Chebyshev's theorem (Tchebychev 1856) for conformal projections appeared in cartography, followed by the amazing Markov projection (Markov 1895).

Developing the theory of functional analysis, Tchebychev (1856) formulated a criterion for the proximity of a certain polynomial to the original function. This criterion was subsequently named after him (see Chap. 2). Applying this criterion to the set of conformal projections, P. Chebyshev formulated the theorem on the best conformal projections, promising to provide a proof in the near future.

However, more serious and complex mathematical problems caused the great mathematician to forget his promise, and the theorem existed without proof for more than 40 years. Only in 1896 did the Ukrainian mathematician and cartographer Grave (1896, 1911) prove Chebyshev's theorem. At the same time, it became clear that the Postel projection is the best of all possible for the area limited by the almucantar,[2] precisely according to Chebyshev's criterion.

The problem of finding the best conical projection according to Chebyshev's criterion for a region bounded by two parallels was solved by Russian mathematician Markov (1895). It turned out that the solution lies outside completely twice differentiable functions. The solution has two discontinuities for the second derivatives of the mapping functions.

The Markov projection is constructed with a method that is currently used to approximate functions with splines. Considering the venerable age of the Markov projection (Markov 1895), and the officially recognized appearance of splines (Schoenberg 1946), this is apparently the first example of the use of a spline in practice. The Markov projection has unique properties, so the last part of this book is devoted to this projection.

The middle of the twentieth century was marked by the rapid development of the theory of optimal control, and with it the solution of various extremal problems. In mathematical cartography, Chebyshev's theorem was extended to a set of orthogonal equidistant along the meridians projections (Novikova 1985). In the same set of projections, the best one according to the Airy criterion was found, and the best Chebyshev projection with the best distortion distribution was identified (Novikova, 1986a; 1986b). Theorems were formulated for ideal projections according to the Airy criterion and for projections close-to-equal-area (Novikova 1986c; 1989) for areas of arbitrary outlines. The best projections for large areas were constructed for the first time (Snyder 1987; Dyer and Snyder 1989). A detailed study of the problem of finding the best projections is presented in (Bugayevskiy and Snyder 1995, pp. 265–278; Grafarend and Krumm 2006, pp. 85–86).

However, despite the development of the theory of the best projections, they have not been widely applied in practice.

There are many both objective and subjective reasons for this. Let us list the main ones.

It turned out that the best projection for one area is not the best for another area. That is, the **best projection is an individual, exclusive object**. This property of the best projections can be easily seen when considering Chebyshev projections,

[2] The almucantar—small circle on the sphere.

i.e., conformal projections constructed in such a way that the linear scale error is minimal (Pavlov 1967; Bugayevskiy and Snyder 1995, pp. 274–278; Pedzich 2005). The procedure for constructing the Chebyshev projection is quite complex. The absence of a simple template prevents the creation of Chebyshev projections from one pattern.

There are other reasons for not using the best projections. One of them is **tradition**. If, by tradition, a certain projection was used for a particular mapping area, then it will continue to be used for that area in the future. An example is the Krovak projection used in the coordinate systems of the Czech Republic (Cesky urad zememericky a katastralní 2020) and Slovakia (Mugnier 2011), as well as the Rosenmund projection in the Swiss coordinate system (Mugnier 2019).

Another reason is **economic**. Certain funds were spent on creating a map in the selected projection. To switch to another projection, cartographers must perform the entire map creation procedure again. This is a complex process that requires substantial material resources. In particular, Timar et al. (2011) indicate that the creation of cadastral maps is an expensive procedure. However, this can also be said of the procedure for creating any map, not just a cadastral map.

The subsequent development of computer technology led to the hope that in the future, when the construction of the best projections becomes a routine operation, they would spread everywhere. However, this did not happen.

As will be shown in Chap. 8, the linear distortions of already known and frequently used projections turn out to be larger than the linear distortions of the best projections for the same territory. However, the distortions are of the same order as the linear distortions of the best projections. The following question arises: why change the well-known and widely used projection to a new, but better one if the linear distortions do not change by two or three times, but only by 0.4%? This question underscores the importance of a comprehensive evaluation of all factors related to the transition to the best projections.

It would seem that the best projections should be most common for large areas, i.e., for small-scale maps. Studies in this area have been carried out (Gilbert 1974; Dyer and Snyder 1989; Kimerling et al. 1995; Canters 2000; Finn et al. 2004; Jenny et al. 2017a, b; Lapaine and Usery 2017; Gyorffy 2018). However here, too, traditions have a great influence.

Gradually, the view of the best projections begins to change, not in the field of small-scale mapping but, on the contrary, in the field of large-scale mapping. The main resource of our planet is land. The main characteristic of a land parcel is its area (Novikova et al. 2023). The standard price of the parcel, the land tax, and the amount of land lease are calculated based on the area. However, the commonly used projections for large-scale maps are conformal projections, i.e., projections that do not distort angles but greatly distort areas.

As Klebanov and Forrai (2010) point out, errors in determining the areas of parcels can lead to a number of claims by users against the state, as the guarantor of their property, if it turns out that the actual area of the parcel is less than stated in the document and the owners or tenants for a number of years paid an inflated amount of tax or rent for this parcel.

There are several ways to solve the problem of area distortion on large-scale maps, which are discussed in detail in Chap. 6.

Chapters 6 and 7 examine in detail the Gauss–Kruger projection; its properties, advantages and disadvantages; and the degree of closeness to the best conformal projection. It will be shown that the conformity of this projection is not an indisputable advantage. Why did the great K. F. Gauss choose the conformal projection, which is named after him, to adjust the Hanoverian triangulation? It is no longer possible to answer this question unambiguously. However, one fairly plausible answer is presented in these chapters. At the same time, two projections are proposed that are quite close in properties to the Gauss–Kruger projection. In the first projection, the area distortions are two times less than the area distortions in the Gauss–Kruger projection. The second projection has practically no area distortions.

In addition to the above chapters, the monograph has five more chapters.

Chapter 1 provides the briefest possible information on the distortion theory developed by Tissot (1881). Some issues of this theory are considered from the perspective of finding the best projections. In particular, we consider such a property of projections as mutual uniqueness, in which only one projection point corresponds to one point on the surface. Not taking this property into account can lead to obtaining a projection that is formally the best, but one in which parts of the image of the original surface overlap in the projection. Such a projection ceases to be one-to-one and is of no interest for practical use. An indicator of unambiguity for the projections is proposed, making it possible to identify projection areas that overlap.

Chapter 2 examines the Euler–Urmaev system of differential equations and provides an answer to the question of why the surfaces of an ellipsoid and a sphere cannot be displayed on a plane without distortion. Additionally, various criteria for ideal projections are analyzed.

Chapter 3 is devoted to the problem of finding an ideal projection using the Airy criterion for an arbitrary mapping area, which is solved by the variational method using the Euler–Ostrogradsky system of equations under natural boundary conditions. After identical transformations, the Euler–Ostrogradsky system is transformed into a system of differential equations for two conjugate harmonic functions. At the same time, on the boundary of the mapping area of the Airy ideal projection, one of the extremal scale factors is equal to unity, and the Tissot's indicatrixes are oriented along tangents or normals to the boundary of the area. The properties of an Airy ideal projection are considered.

In Chap. 4, the same method is applied not to all continuously differentiable projections, but to the set of projections in which the sum of the extremal scale factors is equal to 2. For these projections, the area distortions are quantities of the second order of smallness, whereas the linear distortions are quantities of the first order of smallness. The properties of the resulting projection are considered, and its connection with the best conformal and ideal projections according to the Airy criterion is shown.

Chapter 5 analyzes the Airy projection in detail. It has been proven that the Airy projection is an ideal Airy projection for the area limited by the almucantar. It is shown that the Airy projection, constructed over the entire territory of the globe, loses the property of one-to-one uniqueness, i.e., one projection point corresponds to two surface points. The Airy projection is constructed with maximum coverage of the globe without losing the one-to-one property. This projection covers 98.5% of the globe. However, the minimum linear scale factor of this projection at the boundary is close to zero. Therefore, another Airy projection was built with 93% coverage of the territory, which does not have the disadvantages of the previous version of the projection and can be used to show almost the entire territory of the globe.

Perhaps this book will spark interest in users toward the best projections and inspire further research and practical application of these projections. It is hoped that the presented materials will be useful and engaging for everyone striving for a deep understanding of the problems of mathematical cartography.

References

Airy GB (1861) Explanation of a projection by balance of errors for maps applying to a very large extent of the Earth's surface and comparison of this projection with other projections. London, Edinburgh, Dublin Philosoph Mag 4, 22(149):409–421. 10.1080/14786446108643179

Brown LA (1949) The story of maps. Dover publications Inc., New York, p 464. https://www.goodreads.com/book/show/1901664.The_Story_of_Maps

Bugayevskiy LM, Snyder JP (1995) Map projections. Reference Manual. Taylor&Francis Ltd., London, p 328. https://doi.org/10.1201/b16431

Canters F (2000) Advances in small-scale map projection research. In: Special issue: 29th International geographical congress. In: National committee of geography of Belgium, Société Royale Belge de Géographie, 30 Dec 2000, vol 1, no 2, pp 47–64. 10.4000/belgeo.13883. https://www.researchgate.net/publication/291409904_Advances_in_small-scale_map_projection_research

Caparrini S (2014) The history of the Méchanique analitique. Lett Mat Int 2:47–54. 10.1007/s40329-014-0043-3. https://link.springer.com/article/10.1007/s40329-014-0043-3

Cesky urad zemememericky a katastralní (2020) Analyza stanovení jednotneho referencniho polohoveho a vyskoveho souradnivoveho systenu vcetne zpusobu transformace, p 13. https://www.dia.gov.cz/wp-content/uploads/2023/04/Analyza_referencnich_systemu_pro_NaSaPO.pdf

Dyer JA, Snyder JP (1989) Minimum-error equal-area map projections. American Cartographer 16(1):39–46. https://www.tandfonline.com/doi/abs/10.1559/152304089783875613

Finn MP, Usery EL, Posch ST, Seong JC (2004) A decision support system for map projections of small scale data. Scientific Investigations Report 2004–5297. U.S. Department of the Interior U.S. Geological Survey, p 11. https://permanent.fdlp.gov/LPS102245/carto-research/carto-research.er.usgs.gov/projection/pdf/dss_paper_final.pdf

Gilbert EN (1974) Distortion in maps, review SIAM: society for industrial and applied mathematics 16(1):47–62. https://www.jstor.org/stable/2028761

Grafarend EW, Krumm FW (2006) Map projections. Cartographic information systems. Springer-Verlag Berlin Heidelberg, p 713. https://books.google.com.ua/books/about/Map_Projections.html?id=Rg3Lc8omIXMC&redir_esc=y

Grave DA (1896) Sur la construction des Cartes geographiques Journal de mathematiques pures et appliquees, 5(2):317–361. http://www.numdam.org/item?id=JMPA_1896_5_2__317_0

Gravé D (1911) Demonstration d'un theoreme de Tchebychef generalize. Journal fur die reine und angewandte Mathematik. Band 140. Heft 4, 34:247–251. http://resolver.sub.uni-goettingen.de/purl?PPN243919689_0140

Gyorffy J (2018) Minimum distortion pointed-polar projections for world maps by applying graticule transformation. Int J Cartography 4:2:224–240. 10.1080/23729333.2018.1455263

Jenny B, Savric B (2017a) Combining world map projections. In: Lapaine M, Usery EL (ed) Choosing a map projection, lecture notes in geoinformation and cartography. Springer International Publishing AG, pp 203–211. 10.1007/978-3-319-51835-0_8. https://www.researchgate.net/publication/315860317_Combining_World_Map_Projections

Jenny B, Savric B, Arnold ND, Marston BE, Preppernau CA (2017b) A guide to selecting map projections for world and hemisphere Map. In: Lapaine M, Usery EL (ed) Choosing a map projection, lecture notes in geoinformation and cartography. Springer International Publishing AG, pp 213–228. 10.1007/978-3-319-51835-0_9. https://www.researchgate.net/publication/315857917_A_Guide_to_Selecting_Map_Projections_for_World_and_Hemisphere_Maps

Kimerling AJ, Overton WS, White D (1995) Statistical comparison of map projection distortions within irregular areas. Cartography Geogr Inf Syst 22(3):205–221. 10.1559/152304095782540348

Klebanov M, Forrai J (2010) Implementation of coordinate based Cadastre in Israel: experience and perspectives. In: Paper presented at FIG congress 2010 facing the challenges—building the capacity, Sydney, Australia, 11–16 Apr 2010, p 17. https://www.mapi.gov.il/ProfessionalInfo/MapiPublications/fig2010_klebanov_forrai_3794.pdf

Lapaine M (2016) George Biddell Airy and his contribution to map projections theory. In: Paper presented at the 6th international conference on cartography and GIS, 13–17 June 2016, Albena, Bulgaria ISSN: 1314-0604

Lapaine M, Usery EL (2017) Choosing a map projection. In: Lecture notes in geoinformation and cartography. Springer International Publishing AG, p 357. ISBN 978-3-319-51835-0. https://link.springer.com/book/10.1007/978-3-319-51835-0

Markov AA (1895) O naivygodneyshikh izobrazheniyakh nekotoroy chasti dannoy poverkhnosti vrashcheniya na ploskosti (On the most advantageous images of a certain part of a given surface of rotation on a plane), Izvestiya Imperatorskoy Akademii Nauk 2(3):177–187. https://www.mathnet.ru/rus/im4207

Mugnier CJ (2011) Grids & Datums: Slovak Republic. Photogram Eng Remote Sens 663–664. https://www.asprs.org/wp-content/uploads/2012/05/07-2011-Slovak_Republic.pdf

Mugnier CJ (2019) Grids & Datums: The Swiss Confederation. Photogrammetric Engineering & Remote Sensing, June 2019, 85(6):409–410. 10.14358/PERS.85.6.410

Novikova EN (1985) Rasprostraneniye teoremy Chebysheva na klass ortogonal'nykh ravnopromezhutochnykh vdol' meridianov proyektsiy (Extension of Chebyshev's theorem to the class of orthogonal projections equidistant along meridians). Krivorozh. Gornorudn. in-t. Krivoy Rog, 1985. 17 p.: il. Bibliogr. 11 nazv. Rus. Dep. v UkrNIINTI 12.03.1985, N 512 Uk-95Dep

Novikova EN (1986a) Nailuchshaya ravnopromezhut ochnaya vdol' meridianov proyektsiya variatsionnogo tipa (The best variational type projection equidistant along the meridians). - Krivorozh. Gornorudn. in-t, Krivoy Rog, p 9. Bibliogr. 8 nazv. Rus. Dep. v UkrNIINTI, 02.01.1986, N 49 Uk-86Dep

Novikova EN (1986b) Nailuchshaya po kriteriyu Chebysheva ravnopromezhutochnaya vdol' meridianov proyektsiya s nailuchshim raspredeleniyem iskazheniy (The best projection according to the Chebyshev criterion in equidistant along the meridians with the best distribution of distortions). Krivorozh. Gornorudn. in-t, Krivoy Rog, p 16. Bibliogr. 9 nazv. Rus. Dep. v UkrNIINTI, 09.22.1986, N 2047 Uk-86Dep

Novikova EN (1986c) Problema otyskaniya nailuchshikh proyektsiy v semeystve blizkikh k ravnovelikim (The problem of finding the best projections in a set of close to equal areas). Krivoy Rog, p 21. Bibliogr. 20 nazv. Rus. Dep. v UkrNIINTI, 22.09.1986, N 2071 Uk-86Dep

Novikova EN (1989) Ideal'naya po kriteriyu Eyri i nailuchshaya iz semeystva blizkikh k ravnov elikim proyektsii dlya obshirnykh oblastey proizvol'nykh ochertaniy (Ideal according to the Airy criterion and the best of the family of close-to- equal area projections for large areas of arbitrary shape). Krivoy Rog, p 53. Bibliogr. 14 nazv. Rus. Dep. v UkrNIINTI, 8.06.1989, N 1567 Uk-89Dep

Novikova E, Palamar A, Lopunov R (2023) The dual choice of geodetic horizontal reference systems for Ukraine. Geod Geodyn 14(3):254–264. 10.1016/j.geog.2022.11.001

Pavlov AA (1967) Primeneniye polinomov Chebysheva dlya rascheta kartograficheskoy proyektsii po eskizu setki (Application of Chebyshev polynomials for cartographic projection computation from grid sketches). Novye prob. i metody kartogr, pp 32–39. Nauka, Leningrad

Pedzich P (2005) Conformal projection with minimal distortions. Warsaw University of Technology, Institute of Photogrammetry and Cartography, Politechniki, p 10. https://icaci.org/files/documents/ICC_proceedings/ICC2005/htm/pdf/oral/TEMA2/Session%202/PAWEL%20PEDZICH2.pdf

Schoenberg IJ (1946) Contributions to the problem of approximation of equidistant data by analytic functions: part a—on the problem of smoothing or graduation. A first class of analytic approximation formulae. Q Appl Math 4(1):45–99. https://www.jstor.org/stable/43633538

Snyder JP (1987) Map projectionsa—working manual, U.S. Geological Survey professional paper 1395, USPOGO, Washington D.C., p 397. 10.3133/pp1395

Tchebychev P (1856) Sur la construction des cartes géographiques. Notes 13. (Lu le 18 janvier 1856). Bulletin de la Classe Phisico-Mathématique de L'Académie Impériale des Sciences de Saint-Pétersbourg T. XIV. 17(529):257–261. https://www.biodiversitylibrary.org/item/173056#page/629/mode/1up

Timar G, Baiocchi V, Lelo K (2011) Geodetic Datums of the Italian Cadastral Systems. Geographia Technica, 2011 1:82–90. http://technicalgeography.org/pdf/1_2011/08_gabor_timar_valerio_baio.pdf

Tissot NA (1881) Mémoire sur la représentation des surfaces et les projections des cartes géographiques, Gauthier–Villars, Paris 1881, p 440. https://books.google.com.ua/books/about/Mémoire_sur_la_représentation_des_surf.html?id=eYWxzwEACAAJ&redir_esc=y

Contents

Symbols

The list contains symbols that are widely used. The symbols used in one or two formulae are not specified here.

a	Maximum scale factor at point on the map
b	Minimum scale factor at point on the map
c	Scale distortion
E	Easting axis of the geodetic rectangular coordinate system on the plane
e	(1) the first eccentricity for ellipsoid of revolution; (2) Gaussian coefficient of forward projection (Sect. 1.2)
e'	(1) second eccentricity for ellipsoid of revolution; (2) Gaussian coefficient of inverse projection (Sect. 1.6.5)
f	Gaussian coefficient of forward projection (Sect. 1.2)
f'	Gaussian coefficient of inverse projection (Sect. 1.6.5)
$G(A)$	Green's function for a point A
g	Gaussian coefficient of forward projection (Sect. 1.2)
g'	Gaussian coefficient of inverse projection (Sect. 1.6.5)
h	Differential determinant of Jacoby (Jacobian)
i	(1) angle between the meridian and the parallel of the cartographic grid on the plane (Chap. 1); (2) imaginary unit (Chap. 6)
L	Lagrange multiplier
l	Difference between the longitude of a point and the longitude of the central meridian of the zone
l_0	Difference between the longitude of the boundary meridian and the longitude of the central meridian of the zone
M	Radius of curvature of meridian on ellipsoid of revolution
m	Linear scale factor along meridian
m_0	Linear scale factor along the central meridian of the zone
N	Northing axis of the geodetic rectangular coordinate system on the plane
n	Linear scale factor along parallel
p	Area scale factor
Q	Integrand of the criterion for evaluation of the linear distortions of projection

Q'	Lagrange function
q	Isometric latitude
R	Radius of the sphere of the Earth
r	(1) radius of parallel of ellipsoid of revolution; (2) coefficient of the metric shape of the surface
t	Tangent of geodetic latitude
U	Function relating isometric latitude and geodetic one
V	Linear distortion at an arbitrary point along an arbitrary direction
V_a	Maximum linear distortion
V_b	Minimum linear distortion
V_p	Area distortion
z	Zenith distance of the point on the sphere
A	Grid azimuth of direction on a plane in projection
A_0	Grid azimuth of the main direction on the plane
α	Azimuth of direction on the surface
α'	Image of the azimuth of direction on a plane in projection
α_0	Azimuth of main direction
α_s	Azimuth of tangent to the border of the region
β	Difference between the main direction azimuth on an ellipsoid and the grid azimuth of the main direction on a plane
γ	Grid meridian convergence
Δ	Laplace operator (Chap. 1)
∇	Gradient operator
δ_α	Distortion of the azimuth during the transition from an ellipsoid to a plane in the projection
$\bar{\delta}$	Arc-to-chord correction
ε	Distortion of the angle between the meridian and the parallel of the cartographic grid on the plane
ε_J	General Jordan measure of linear distortion at a point
ε_A	General Airy measure of linear distortion at a point
ε_{Ch}	General Chebyshev measure of linear distortion at a point
ε_M	General Markov measure of linear distortion at a point
ζ	Azimuth of the direction of the gradient of the Green's function
η	(1) axis of an elementary rectangular coordinate system on the surface of an ellipsoid (Chap. 1); (2) auxiliary quantity used in determining the mapping functions of the Gauss–Kruger projection, namely, $= \sqrt{e'^2\cos^2\varphi}$ (Chap. 6)
ϑ	Axis of the geodetic polar coordinate system on the surface of an ellipsoid
κ	Constant parameter of conic and cylindrical projections
λ	Geodetic longitude
μ	Linear scale factor of forward projection
μ'	Linear scale factor of inverse projection
μ_0	Nominal linear scale
ρ	Radius vector point on plane in projection of polar coordinate system
σ	Mapping area
$\bar{\sigma}$	Indicator of unambiguity

Ξ	Criterion for evaluation of the linear distortions of projection
τ	Natural logarithm of coefficient of metric shape of surface
ξ	(1) axis of an elementary rectangular coordinate system on the surface of an ellipsoid (Chap. 1); (2) axis of a geodetic polar coordinate system on the surface of an ellipsoid (Chap. 8)
Φ	Harmonic function conjugate to the function β for an ideal Airy projection
φ	Geodetic latitude
χ	Sum of the main direction azimuth on an ellipsoid and the grid azimuth of the main direction on a plane
Ω	Coefficient of metric shape of surface of geodetic polar coordinate system
ω	Maximum angular distortion

Chapter 1
Map Projections and Their Distortions

The inconstancy and nonmathematical nature of the physical surface of the Earth have led to the fact that the physical surface is projected onto the surface of a reference ellipsoid or sphere, and then, this surface is projected onto a plane. Let us consider the procedure for projecting the surface of an ellipsoid of revolution and a sphere onto a plane in more detail.

1.1 Basic Formulae and Definitions

In accordance with (Bronstejn et al. 2020), a surface S is mapped onto a surface S' if every point A on the surface S corresponds to a certain point A' on the surface S', as shown in Fig. 1.1.

For cartography, the first surface is the surface of an ellipsoid of revolution or a sphere, and the second surface is a plane.

The surface of the sphere is determined by one parameter, the radius of the sphere, R.

The ellipsoid of revolution is determined by two parameters: the semi-major axis and the flattening of the ellipsoid (NGA.STND.0036_1.0.0_WGS84 2014).

Both the sphere and the Earth's ellipsoid are often called the **mathematical surface of the Earth**, since in both cases, the Earth can be represented as a fairly simple regular surface. The positions of the points on these surfaces can be obtained by using geodetic coordinates:φ—latitude; and λ—longitude.

Traditionally, until the end of the twentieth century in Europe, a version of the flat Cartesian coordinate system, called left-handed or geodetic, was used to solve problems of geodesy and cartography on a plane. In this version (see Fig. 1.2a), the X-axis is directed upward, and the Y-axis is directed to the right. The direction of the counting angles is clockwise from the X-axis, as shown in Fig. 1.2a.

E. Novikova, *Best Map Projections*, https://doi.org/10.1007/978-3-031-78334-0_1

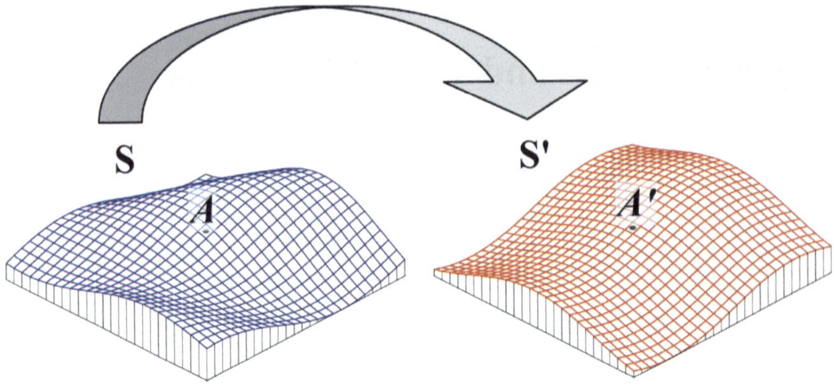

Fig. 1.1 Visual representation of the process of mapping one surface (S) onto another (S')

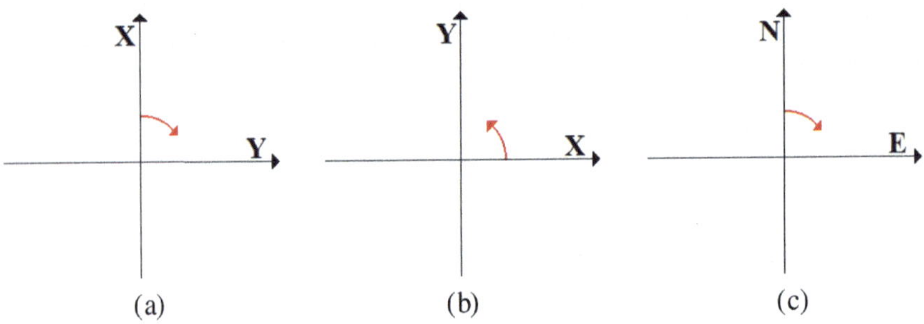

Fig. 1.2 Variants of the rectangular coordinate system: **a** geodetic left-handed (obsolete), and **b** mathematical right-handed; **c** geodetic left-handed (modern)

At the same time, surveyors and cartographers in America use a flat rectangular coordinate system, which is accepted in mathematics and called right-handed (Bugayevskiy and Snyder 1995). In this system, the X-axis is directed to the right, and the Y-axis is directed upward. The direction of the counting angles is counterclockwise, as shown in Fig. 1.2b.

These two systems have both advantages and disadvantages.

The advantage of the left-handed system is that when the surface of the Earth's ellipsoid is projected onto a plane, the X-axis almost coincides with the direction of the meridian. The direction of the counting angles on the plane coincides with the direction of the counting azimuths on the Earth's surface.

The advantage of the right-hand system is that it completely coincides with the rectangular coordinate system on the plane accepted in mathematics. This means that when solving geometric problems in this system, we can use formulae and algorithms of classical mathematics.

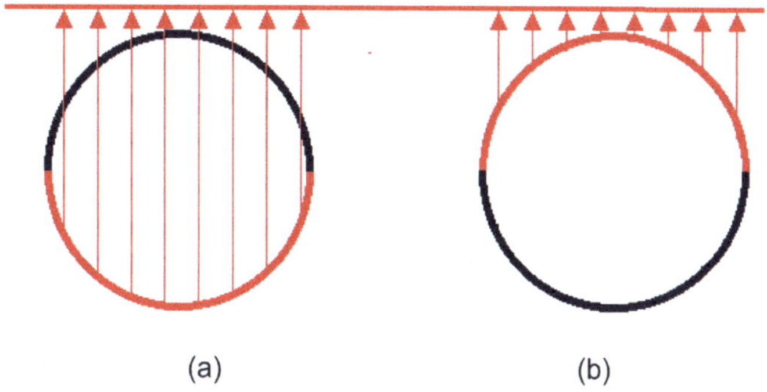

Fig. 1.3 Design procedures for **a** positive projections, and **b** negative projections

The measurement of angles in both systems is different. In the left-hand system, angles are counted clockwise; in the right-hand system, they are counted counterclockwise.

The disadvantage of the right-hand system is that its use in cartography leads to a change in the signs of some quantities. In particular, the signs of the partial derivatives of the mapping functions and quantities dependent on them change. One of these quantities is the Jacobian of the projection (see Sect. 1.6), which is part of the formula for calculating the area scale factor. In traditional mathematical cartography, the Jacobian is positive for positive projections and negative for negative projections. The difference between positive and negative projections is clearly shown in Fig. 1.3. The mapping process is shown with red arrows. The displayed surface is also colored red. For positive projections, mapping is performed such that the observer is above the Earth's surface. For negative projections, the mapping is performed as if from inside the Earth. The concept of positive and negative projections is introduced as a rule for perspective azimuthal projections. However, if the sign of the Jacobian of the projection is used as an indicator of the positivity property, then such a definition can be extended to all possible map projections.

When transitioning from a right-handed coordinate system to a left-handed system, it is necessary to modify the formulae for determining values that should be positive by default to preserve their signs. Additionally, differences in the directions of angle measurements must be considered. This procedure for transitioning from a right-handed coordinate system to a left-handed system was performed by Bugayevskiy and Snyder (1995).

The development and spread of unified geodetic software forced the geodetic community to seek a compromise between left-handed and right-handed coordinate systems to eliminate confusion between them. This compromise was found by introducing a coordinate system in which the axes had names different from those previously accepted. The axis N (Northing axis) is directed upward, and the axis E (Easting axis) is directed to the right. Angles are measured from the axis N clockwise (see Fig. 1.2c), i.e., the system is left-handed.

The N, E coordinate system in mathematical cartography has not yet been widely used. However, this is the system used in this work. This choice allows the use of traditional formulae of mathematical cartography without any changes. To do this, in the formulae of mathematical cartography, the following substitutions are sufficient:

$$\mathbf{X} \rightarrow \mathbf{N}$$
$$\mathbf{Y} \rightarrow \mathbf{E}$$

Moreover, the new system avoids problems associated with the transition to a mathematical coordinate system. To display the resulting projections using graphic programs with a mathematical coordinate system, it is sufficient to replace the N, E coordinates with the X, Y coordinates of the mathematical coordinate system according to the following scheme:

$$\mathbf{N} \rightarrow \mathbf{Y}$$
$$\mathbf{E} \rightarrow \mathbf{X}$$

In geodesy, the names of the axes Northing and Easting are justified by the fact that the axes actually have a direction to the North and East, respectively. In mathematical cartography, geometric correspondence can be violated since the North Pole can be located at any point in the projection. However, taking into account the above-described advantages of the N, E system, this particular system of rectangular coordinates on the plane is adopted in this work. In this case, a point on the surface of an ellipsoid or a sphere with coordinates φ, λ corresponds to a point with coordinates N, E on the plane. The mapping process can be written mathematically as follows:

$$N = f_1(\varphi, \lambda), E = f_2(\varphi, \lambda). \tag{1.1}$$

The formulae (1.1) are called **map projection equations**. The functions $f_1(\varphi, \lambda), f_2(\varphi, \lambda)$ are called **mapping functions**.

There are two restrictions of mapping functions:

1. **The functions must be one-to-one.** In this case, one projection point corresponds to only one point on the original surface. Similarly, one point on the original surface corresponds to only one projection point.
2. **The functions must be continuously differentiable everywhere.**

These conditions in the theory of best projections are contradictory. As shown below, solving the extremal problem of finding the best projection without taking into account the principle of one-to-one can lead to a solution that is formally the best but in which parts of the image of the original surface in the projection are superimposed on one another (see Sect. 5.5). On the other hand, taking into account the principle of one-to-one uniqueness can lead to the solution of the problem of finding the best projection being outside the set of everywhere twice continuously differentiable functions (see Sect. 8.3).

Lines of equal coordinates on the surface of an ellipsoid are lines of meridians and parallels. Their image on a projection plane is called a **cartographic grid**.

The equation of parallels on the surface of an ellipsoid looks like this:

$$\varphi = \varphi_0 = \text{const.} \tag{1.2}$$

On a plane, the equations of parallels are written as follows:

$$N = f_1(\varphi_0, \lambda), E = f_2(\varphi_0, \lambda). \tag{1.3}$$

Similarly, the equations of the meridians on a plane can be written as:

$$N = f_1(\varphi, \lambda_0), \quad E = f_2(\varphi, \lambda_0). \tag{1.4}$$

Equations (1.3) and (1.4) are called equations of parallels and meridians in parametric form.

Equations (1.3) can be considered a system of two equations with three unknowns (N, E, λ). One unknown can be eliminated. In this case, one equation with two unknowns is obtained. If we transform system (1.3) in this way, we obtain:

$$F_1(N, E, \varphi_0) = 0. \tag{1.5}$$

A similar transformation of system (1.4) yields the following equation:

$$F_2(N, E, \lambda_0) = 0. \tag{1.6}$$

Equations (1.5) and (1.6) are called the equations of parallels and meridians in canonical form.

To solve the problem of finding the best projections, the most convenient is an isometric coordinate system, q, λ, where λ is the geodetic longitude and where q is the isometric latitude, related to geodetic latitude by the known equation (Bugayevskiy and Snyder 1995, p. 12):

$$\mathrm{d}q = \frac{M}{r}\mathrm{d}\varphi, \tag{1.7}$$

where M, r are the radii of curvature of meridians and radii of parallels, equal to (Rapp 1991, p. 30, 35):

$$M = \frac{a\left(1 - e^2\right)}{(1 - e^2 \sin^2 \varphi)^{\frac{3}{2}}}, r = \frac{a \cos \varphi}{(1 - e^2 \sin^2 \varphi)^{\frac{1}{2}}}. \tag{1.8}$$

In formulae (1.8), a is the semi-major axis of the ellipsoid; e is the first eccentricity for the ellipsoid of revolution.

By integrating formula (1.7) under the condition that at the equator, when $\varphi = 0$, the isometric latitude is also equal to zero, we obtain:

$$q = \ln\left(\tan\left(\frac{\pi}{4} + \frac{\varphi}{2}\right)\left(\frac{1 - e\sin\varphi}{1 + e\sin\varphi}\right)^{\frac{e}{2}}\right). \tag{1.9}$$

The use of notation (Bugayevskiy and Snyder 1995, p. 12):

$$U = \tan\left(\frac{\pi}{4} + \frac{\varphi}{2}\right)\left(\frac{1 - e\sin\varphi}{1 + e\sin\varphi}\right)^{\frac{e}{2}}, \tag{1.10}$$

allows us to rewrite Eq. (1.9) in the following form:

$$q = \ln U. \tag{1.11}$$

If the initial surface is the surface of the sphere ($e = 0$), then:

$$q = \ln\left(\tan\left(\frac{\pi}{4} + \frac{\varphi}{2}\right)\right). \tag{1.12}$$

To study the mapping functions, the behavior of their derivatives is examined. The following notations for partial derivatives are used in cartography:

$$\frac{\partial N}{\partial q} = N_q, \frac{\partial N}{\partial \lambda} = N_\lambda, \frac{\partial E}{\partial q} = E_q, \frac{\partial E}{\partial \lambda} = E_\lambda. \tag{1.13}$$

There are combinations of derivatives used in differential geometry, such as:

$$e = N_q^2 + E_q^2, g = N_\lambda^2 + E_\lambda^2, f = N_q N_\lambda + E_q E_\lambda, h = N_q E_\lambda - N_\lambda E_q. \tag{1.14}$$

The quantities e, g, f are called coefficients of the first fundamental form of the surface, or Gaussian coefficients, and h is the differential determinant of Jacoby, or Jacobian.

Only three of the four quantities are independent. Therefore, they are related by an equation called the **Euler–Lagrange identity**, namely:

$$f^2 + h^2 = \text{eg}. \tag{1.15}$$

Two quantities, e, g, are always nonnegative.

Let us write the formulae for the total differentials of mapping functions:

$$\mathrm{d}N = N_q \mathrm{d}q + N_\lambda \mathrm{d}\lambda, \ \mathrm{d}E = E_q \mathrm{d}q + E_\lambda \mathrm{d}\lambda. \tag{1.16}$$

1.2 Relationship Between Spheroidal and Planar Azimuths

Let us take a point A with coordinates φ, λ on the surface of the ellipsoid and an infinitely close point B with coordinates $\varphi + \mathrm{d}\varphi$, $\lambda + \mathrm{d}\lambda$. Let us drop a perpendicular from point B to the meridian of point A. In the resulting right triangle, the angle ∠CAB is the spheroidal azimuth (Fig. 1.4a).

The arc element on an ellipsoid along the meridian line, as stated in (Rapp 1991, p. 30), is equal to:

$$\mathrm{d}S_{\mathrm{mer}} = M\,\mathrm{d}\varphi = r\mathrm{d}q. \tag{1.17}$$

Similarly, an element on an ellipsoid along a line of parallel is equal to:

$$\mathrm{d}S_{\mathrm{par}} = r\mathrm{d}\lambda. \tag{1.18}$$

From the right triangle ABC, we write:

$$\tan\alpha = \frac{r\mathrm{d}\lambda}{r\mathrm{d}q} = \frac{\mathrm{d}\lambda}{\mathrm{d}q}. \tag{1.19}$$

On a plane, point A will be represented by point A' with coordinates N, E, and point B will be represented by point B' with coordinates $N+\mathrm{dN}$, $E+\mathrm{dE}$. A spheroidal azimuth α is represented by a flat azimuth α'. On a plane, angles are measured not from the image of the meridian but from the axis N. These angles are called the grid azimuth. Let us write the equation for the grid azimuth of direction $A'B'$ (Fig. 1.4b).

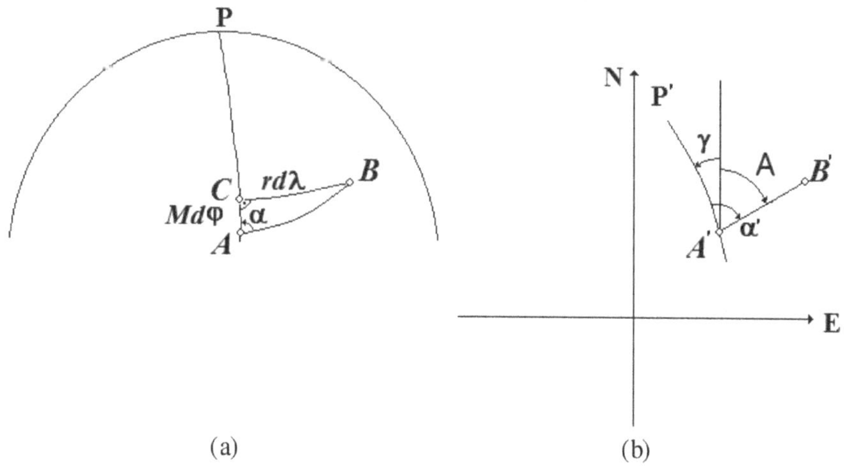

Fig. 1.4 Relationship between **a** spheroidal, and **b** planar azimuths

$$\tan A = \frac{\mathrm{d}E}{\mathrm{d}N} = \frac{E_q \mathrm{d}q + E_\lambda \mathrm{d}\lambda}{N_q \mathrm{d}q + N_\lambda \mathrm{d}\lambda}. \tag{1.20}$$

For a flat azimuth, we can write:

$$\alpha' = A + \gamma. \tag{1.21}$$

In formula (1.21), γ is the grid meridian convergence. If the point is taken on the meridian, then the flat azimuth is zero. Therefore, the angle γ can be considered the grid azimuth of the meridian, counted in the opposite direction, i.e.,

$$\gamma = -A|_{\substack{\alpha=0, \\ d\lambda=0}}. \tag{1.22}$$

Thus:

$$\tan\gamma = -\frac{E_q}{N_q}. \tag{1.23}$$

Using formula (1.21), we write:

$$\tan\alpha' = \tan(A+\gamma) = \frac{\tan A + \tan\gamma}{1 - \tan A \tan\gamma}. \tag{1.24}$$

Substituting (1.20, 1.23) into (1.24) leads to:

$$\tan\alpha' = \frac{\frac{E_q \mathrm{d}q + E_\lambda \mathrm{d}\lambda}{N_q \mathrm{d}q + N_\lambda \mathrm{d}\lambda} - \frac{E_q}{N_q}}{1 + \frac{E_q \mathrm{d}q + E_\lambda \mathrm{d}\lambda}{N_q \mathrm{d}q + N_\lambda \mathrm{d}\lambda} \cdot \frac{E_q}{N_q}},$$

or, after identical transformations:

$$\tan\alpha' = \frac{\left(N_q E_\lambda - E_q N_\lambda\right)\mathrm{d}\lambda}{\left(N_q^2 + E_q^2\right)\mathrm{d}q + \left(N_q N_\lambda + E_q E_\lambda\right)\mathrm{d}\lambda}. \tag{1.25}$$

Taking into account formulae (1.14), we write expression (1.25) in the following form:

$$\tan\alpha' = \frac{h\mathrm{d}\lambda}{e\mathrm{d}q + f\mathrm{d}\lambda}, \tag{1.26}$$

or, given (1.19):

$$\tan\alpha' = \frac{h\sin\alpha}{e\cos\alpha + f\sin\alpha}. \tag{1.27}$$

Equation (1.27) is the formula for connecting the spheroidal azimuth with its image on a plane.

Let us denote:

$$\delta_\alpha = \alpha' - \alpha. \tag{1.28}$$

Obviously, δ_α is a distortion of the azimuth during the transition from an ellipsoid to a plane in the projection. Let us find the tangent of this angle:

$$\tan\delta_\alpha = \tan\left(\alpha' - \alpha\right) = \frac{\tan\alpha' - \tan\alpha}{1 + \tan\alpha'\tan\alpha}, \tag{1.29}$$

or, using (1.27),

$$\tan\delta_\alpha = \tan(A + \gamma - \alpha) = \frac{(h - e)\sin\alpha\cos\alpha - f\sin^2\alpha}{e\cos^2\alpha + f\sin\alpha\cos\alpha + h\sin^2\alpha}. \tag{1.30}$$

Combining Eqs. (1.21) and (1.28), we obtain the following formula for the transition from the azimuth on the ellipsoid to the grid azimuth on the plane (Bagratuni 1962, p. 229):

$$A = \alpha - \gamma + \delta_\alpha. \tag{1.31}$$

Formula (1.31) can be considered a formula for the relationship of the grid azimuth on the plane with the azimuth on the ellipsoid.

1.3 Angle Between Meridian and Parallel of Cartographic Grid

Let us denote the angle between the meridian and the parallel of the cartographic grid by i. This angle can be considered a plane azimuth of the parallel, i.e.,

$$i = \alpha'|_{\alpha=90^\circ}. \tag{1.32}$$

Substituting (1.32) into (1.27) leads to:

$$\tan i = \frac{h}{f}. \tag{1.33}$$

Using elementary trigonometric transformations and taking into account the Euler–Lagrange identity (1.15), we obtain (Bugayevskiy and Snyder 1995, p. 16):

$$\sin i = \frac{h}{\sqrt{eg}},\ \cos i = \frac{f}{\sqrt{eg}}. \tag{1.34}$$

The angle between the meridian and the parallel on the surface of the ellipsoid is 90°. On a plane, this angle, according to the formulae (1.33–1.34), may differ from 90°. Let us denote ε by the distortion of the angle between the meridian and the parallel of the cartographic grid, namely:

$$\varepsilon = i - 90. \tag{1.35}$$

Using formula (1.33) yields:

$$\tan \varepsilon = \tan(i - 90°) = -\cot i = -\frac{f}{h}. \tag{1.36}$$

The condition for the orthogonality of the cartographic grid is given below (Bugayevskiy and Snyder 1995, p. 16):

$$\varepsilon = 0, \tag{1.37}$$

or, taking into account (1.36),

$$f = 0. \tag{1.38}$$

1.4 Linear Element on Ellipsoid and in Projection

Considering Eqs. (1.17) and (1.18), the linear element between points A and B on the ellipsoid's surface can be written as follows:

$$\mathrm{d}s^2 = r^2\left(\mathrm{d}q^2 + \mathrm{d}\lambda^2\right), \tag{1.39}$$

or, according to (1.19):

$$\mathrm{d}s = \frac{r\mathrm{d}q}{\cos \alpha}. \tag{1.40}$$

In formulae (1.39–1.40), the function r is called the **coefficient of the metric shape of the surface**. For an isometric coordinate system q, λ, this coefficient is equal to the radius of the parallel. For more details on the surface metric shape coefficient, see Sect. 1.12.

Let us write the linear element on the plane between the points A' and B'.

$$ds'^2 = dN^2 + dE^2. \tag{1.41}$$

Using formulae (1.16), we obtain:

$$ds'^2 = dN^2 + dE^2 = (N_q dq + N_\lambda d\lambda)^2 + (E_q dq + E_\lambda d\lambda)^2, \tag{1.42}$$

or,

$$ds'^2 = \left(N_q^2 + E_q^2\right)dq^2 + 2\left(N_q N_\lambda + E_q E_\lambda\right)dq d\lambda + \left(E_q^2 + E_\lambda^2\right)d\lambda^2. \tag{1.43}$$

By incorporating notation (1.14), formula (1.43) can be rewritten as:

$$ds'^2 = e dq^2 + 2f dq d\lambda + g d\lambda^2. \tag{1.44}$$

Formula (1.44) is called the **formula of the first fundamental form of the surface** or the **Gauss formula** (Bugayevskiy and Snyder 1995, p. 16).

1.5 Curvature of the Image of a Geodesic Line on a Plane

The grid azimuth defined by formula (1.31) is the instantaneous grid azimuth of some point on the curve, i.e., the angle between the direction of the N-axis and the tangent to the curve at a point. When moving from one point of the curve to another, as shown in Fig. 1.5, the value of the grid azimuth changes.

When transferred from an ellipsoid to a plane, a geodesic line is generally depicted as a curve, which is different from a geodesic line on a plane, i.e., from a straight line. Obviously, for any curve, the grid azimuth of a geodesic line is a variable value

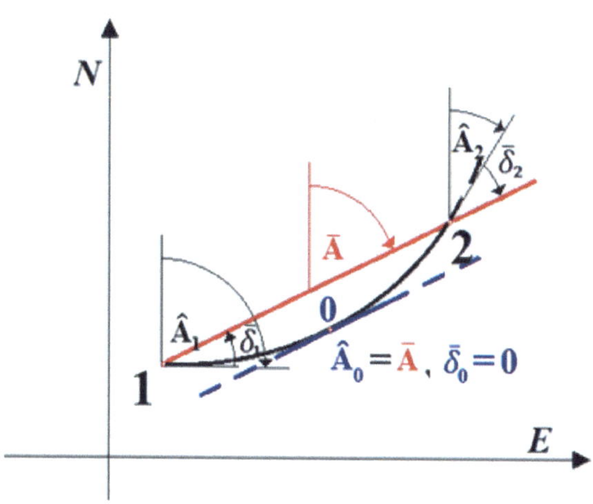

Fig. 1.5 Image of a geodesic (black) and a chord (red) between points 1 and 2 in projection

that is different at each point from the grid azimuth of the chord between the first and end points of the geodesic line (see Fig. 1.5). In this context, the problem arises of transitioning from the grid azimuths of a geodesic line to the grid azimuth of a chord.

The connection between the grid azimuth of the chord and the grid azimuths of the image of the geodesic line at the initial (1) and final (2) points can be written as follows:

$$\overline{A} = \widehat{A}_j + \overline{\delta}_j, (j = 1, 2). \quad (1.45)$$

According to Fig. 1.5, the corrections $\overline{\delta}_1$ and $\overline{\delta}_2$ have different signs.

In Eq. (1.45), $\overline{A}$ is the grid azimuth of the chord between the starting and ending points of the geodesic; $\overline{\delta}_j$ is the arc-to-chord correction; and $\widehat{A}_j$ is the grid azimuth of the line point, which is the image of the geodesic (see Fig. 1.5). At the starting point $\widehat{A}_j = \widehat{A}_1$, at the end point $\widehat{A}_j = \widehat{A}_2$.

Let us rewrite Eq. (1.45) in the following form:

$$\overline{\delta}_j = -\widehat{A}_j + \overline{A}, \quad (1.46)$$

where $\overline{\delta}_j, \widehat{A}_j$ are variables that depend on the distance s from the starting point to the current point of the geodesic line and where $\overline{A}$ is a constant.

We look for the value $\overline{\delta}_j$ as a function of the distance s from the initial point to point j. Using the first terms of the Maclaurin series for $\overline{\delta}_j$, we write:

$$\overline{\delta}_j = \overline{\delta}_1 + \left(\frac{\mathrm{d}\overline{\delta}}{\mathrm{d}s}\right)_1 s + \frac{1}{2}\left(\frac{\mathrm{d}^2\overline{\delta}}{\mathrm{d}s^2}\right)_1 s^2 + \ldots \quad (1.47)$$

Using (1.46) gives:

$$\overline{\delta}_j = \overline{\delta}_1 - \left(\frac{\mathrm{d}\widehat{A}}{\mathrm{d}s}\right)_1 s - \frac{1}{2}\left(\frac{\mathrm{d}^2\widehat{A}}{\mathrm{d}s^2}\right)_1 s^2 + \ldots \quad (1.48)$$

For $s = 0$, Eq. (1.48) gives $\overline{\delta}_1$; for $s = s_{12}$, Eq. (1.48) gives the correction at the end point of the geodesic line, i.e.,

$$\overline{\delta}_2 = \overline{\delta}_1 - \left(\frac{\mathrm{d}\widehat{A}}{\mathrm{d}s}\right)_1 s_{12} - \frac{1}{2}\left(\frac{\mathrm{d}^2\widehat{A}}{\mathrm{d}s^2}\right)_1 s_{12}^2 + \ldots \quad (1.49)$$

Using Eq. (1.49), it is possible to calculate only the difference between the corrections: $\overline{\delta}_2 - \overline{\delta}_1$. To obtain both corrections $\overline{\delta}_1, \overline{\delta}_2$, or, equivalently, to determine the function $\overline{\delta}_j$ at each point of the geodesic, it is necessary to know the value of $\overline{\delta}_i$ at least at one point.

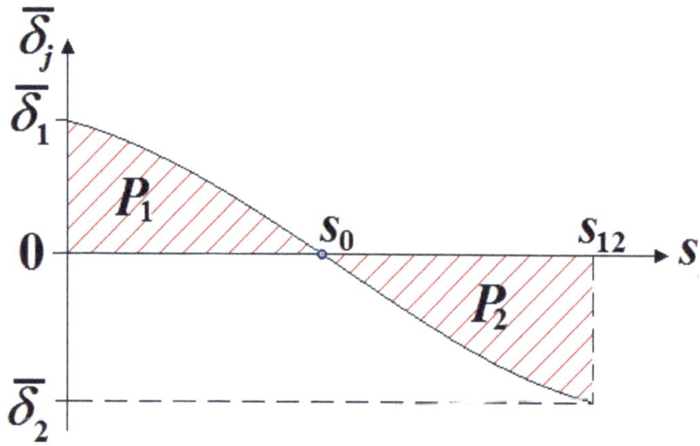

Fig. 1.6 Graph of changes in the function δ_i depending on the distance s

As mentioned earlier, the corrections $\overline{\delta}_1$, $\overline{\delta}_2$ have different signs. This means that there is a point on the geodesic line at which $\overline{\delta}_j$ is equal to zero. In Fig. 1.5, this point is shown in blue. At this point, the grid azimuth of the image of geodesic line is equal to the grid azimuth of the chord between points 1 and 2. Let us denote the distance from the starting point to this point by s_0. Figure 1.6 shows a graph of changes in the function $\overline{\delta}_j$, depending on the distance s along the geodesic, provided that $\overline{\delta}_1 > 0$.

Geodesics have several extreme properties; in particular, the distance between two points along a geodesic is minimal (Bagratuni 1962; Martínez-Llario et al. 2021; Seif and Ghalenoei 2018). However, geodesics have properties related to certain areas. For example, the positive part of the area of the graph $\overline{\delta}_i$ is equal to the negative part, namely:

$$P_1 = -P_2. \tag{1.50}$$

Using Eq. (1.48), we obtain the area P_1:

$$P_1 = \int_0^{s_0} \overline{\delta}_i \mathrm{d}s = \int_0^{s_0} \left(\overline{\delta}_1 - \left(\frac{\mathrm{d}\widehat{A}}{\mathrm{d}s} \right)_1 s - \frac{1}{2} \left(\frac{\mathrm{d}^2\widehat{A}}{\mathrm{d}s^2} \right)_1 s^2 \right) \mathrm{d}s, \tag{1.51}$$

or,

$$P_1 = \overline{\delta}_1 s_0 - \left(\frac{\mathrm{d}\widehat{A}}{\mathrm{d}s} \right)_1 \frac{s_0^2}{2} - \frac{1}{2} \left(\frac{\mathrm{d}^2\widehat{A}}{\mathrm{d}s^2} \right)_1 \frac{s_0^3}{3}. \tag{1.52}$$

Likewise, for P_2:

$$P_2 = \overline{\delta}_1(s_{12} - s_0) - \left(\frac{\mathrm{d}\widehat{A}}{\mathrm{d}s}\right)_1 \left(\frac{s_{12}^2}{2} - \frac{s_0^2}{2}\right) - \frac{1}{2}\left(\frac{\mathrm{d}^2\widehat{A}}{\mathrm{d}s^2}\right)_1 \left(\frac{s_{12}^3}{3} - \frac{s_0^3}{3}\right). \tag{1.53}$$

Substituting (1.52–1.53) into (1.50), after identical transformations and taking into account (1.49), we have the following equations for determining $\overline{\delta}_1, \overline{\delta}_2$:

$$\overline{\delta}_1 = \frac{1}{2}\left(\frac{\mathrm{d}\widehat{A}}{\mathrm{d}s}\right)_1 s_{12} + \frac{1}{6}\left(\frac{\mathrm{d}^2\widehat{A}}{\mathrm{d}s^2}\right)_1 s_{12}^2 .., \overline{\delta}_2 = -\frac{1}{2}\left(\frac{\mathrm{d}\widehat{A}}{\mathrm{d}s}\right)_1 s_{12} - \frac{1}{3}\left(\frac{\mathrm{d}^2\widehat{A}}{\mathrm{d}s^2}\right)_1 s_{12}^2 \ldots \tag{1.54}$$

There are at least two different ways to define the function $\widehat{A}$ and its derivatives. The first way is to use equality (1.31). The second way is to use formula (1.20) for the grid azimuth of the geodesic line.

The first method is simpler but requires knowledge of two functions γ, δ_α and their derivatives. In Sect. 1.7, this method is used to determine the arc-to-chord corrections of the Gauss–Kruger projection. However, the functions γ, δ_α are not always known with sufficient accuracy. Therefore, let us consider the definition of $\widehat{A}$ using Eq. (1.20), namely:

$$\tan \widehat{A} = \frac{E_q \mathrm{d}q + E_\lambda \mathrm{d}\lambda}{N_q \mathrm{d}q + N_\lambda \mathrm{d}\lambda}. \tag{1.55}$$

Using formulae (1.19, 1.39), we can find the sine and cosine of the azimuth along direction ds on the ellipsoid:

$$\sin \alpha = r\frac{\mathrm{d}\lambda}{\mathrm{d}s}, \tag{1.56}$$

$$\cos \alpha = r\frac{\mathrm{d}q}{\mathrm{d}s}. \tag{1.57}$$

Considering (1.56–1.57), Eq. (1.55) is written as follows:

$$\tan \hat{\mathrm{A}} = \frac{E_q \cos \alpha + E_\lambda \sin \alpha}{N_q \cos \alpha + N_\lambda \sin \alpha}. \tag{1.58}$$

Since the variable s depends on two other variables q, λ, the derivative of $\widehat{A}$ with respect to s is equal to:

$$\frac{\mathrm{d}\widehat{A}}{\mathrm{d}s} = \frac{\partial \widehat{A}}{\mathrm{d}q} \cdot \frac{\mathrm{d}q}{\mathrm{d}s} + \frac{\partial \widehat{A}}{\partial \lambda} \cdot \frac{\mathrm{d}\lambda}{\mathrm{d}s}. \tag{1.59}$$

Let us denote the numerator of Eq. (1.58):

$$C = E_q \cos\alpha + E_\lambda \sin\alpha, \tag{1.60}$$

and find its derivatives:

$$C_q = E_{qq} \cos\alpha + E_{q\lambda} \sin\alpha + \left(-E_q \sin\alpha + E_\lambda \cos\alpha\right)\alpha_q, \tag{1.61}$$

$$C_\lambda = E_{q\lambda} \cos\alpha + E_{\lambda\lambda} \sin\alpha + \left(-E_q \sin\alpha + E_\lambda \cos\alpha\right)\alpha_\lambda. \tag{1.62}$$

Similar derivatives also occur for the denominator of Eq. (1.58).

To determine the derivatives α_q, α_λ included in formulae (1.61–1.62), it is important to consider that the azimuth of the direction on the ellipsoid is the azimuth of the geodesic that satisfies the Clairaut equation (Bagratuni 1962; Sjöberg 2012):

$$r \cdot \sin\alpha = \text{const.} \tag{1.63}$$

Considering that r depends only on latitude, we obtain:

$$\alpha_q = -\tau_q \tan\alpha,\ \alpha_\lambda = 0, \tag{1.64}$$

where τ_q is the derivative with respect to q of the function:

$$\tau = \ln r. \tag{1.65}$$

The formulae for calculating τ_q and other derivatives of τ are presented in Appendix A.

By differentiating equality (1.58) with respect to q, λ and substituting the resulting derivatives into (1.59) after identical transformations, taking into account (1.61–1.64), it is possible to obtain:

$$\widehat{A}_s = \frac{1}{r \cdot \mathrm{d}s'^2}\left(d_1 \cos^3\alpha + d_2 \sin\alpha \cos^2\alpha + d_3 \sin^2\alpha \cos\alpha + d_4 \sin^3\alpha - h\tau_q \sin\alpha\right), \tag{1.66}$$

where

$$d_1 = N_q E_{qq} - E_q N_{qq},\ d_2 = 2N_q E_{q\lambda} - 2E_q N_{q\lambda} + N_\lambda E_{qq} - E_\lambda N_{qq}, \tag{1.67}$$

$$d_3 = 2N_\lambda E_{q\lambda} - 2E_\lambda N_{q\lambda} + N_q E_{\lambda\lambda} - E_q N_{\lambda\lambda},\ d_4 = N_\lambda E_{\lambda\lambda} - E_\lambda N_{\lambda\lambda}. \tag{1.68}$$

In Eq. (1.66), h, $\mathrm{d}s'^2$ are the Jacobian of the projection and the square of the linear element in the projection, determined by formulae (1.14) and (1.44), respectively.

Repeatedly differentiating the derivative of $\widehat{A}_s$ with respect to s gives:

$$\widehat{A}_{ss} = \left(\widehat{A}_s\right)_q \cdot \frac{1}{r}\cos\alpha + \left(\widehat{A}_s\right)_\lambda \cdot \frac{1}{r}\sin\alpha. \tag{1.69}$$

Differentiation of Eq. (1.66) with respect to q and λ, substitution of the resulting derivatives into (1.69), and the identical transformation yield the following formula for calculating $\widehat{A}_{ss}$:

$$\begin{aligned}\widehat{A}_{ss} = &\frac{1}{r^2 \mathrm{d}\dot{s}'^2}\left(D_1\cos^4\alpha + D_2\sin\alpha\cos^3\alpha + D_3\sin^2\alpha\cos^2\alpha\right.\\&\left.+D_4\sin^3\alpha\cos\alpha + D_5\sin^4\alpha - H_1\right)\\&-\frac{1}{rds'^2}\widehat{A}_s\left(D_6\cos^3\alpha + D_7\sin\alpha\cos^2\alpha + D_8\sin^2\alpha\cos\alpha + D_9\sin^3\alpha\right).\end{aligned} \tag{1.70}$$

The formulae for calculating the coefficients $D_1, D_2, \ldots D_9, H_1$ are presented in Appendix B.

Equations (1.54, 1.66–1.70) allow us to solve the problem of determining the arc-to-chord corrections for the transition from the grid azimuth of the image of the geodesic line to the grid azimuth of the chord passing through the ends of this line. In Sect. 1.7, using these equations, the arc-to-chord correction is determined for projections that are ideal according to Airy and close-to-equal-area, having properties similar to those of the Gauss–Kruger projection.

1.6 Local Linear Scale

1.6.1 General Formula for the Local Linear Scale

Let us compare the linear elements on an ellipsoid and on a plane. To do this, we introduce the concept of a linear scale.

The local linear scale (Bugayevskiy and Snyder 1995, p. 17) is the limit of the ratio of the length of a segment in a projection on a plane to the length of the corresponding segment on an ellipsoid, as the latter tends to zero:

$$\mu = \lim_{AB\to 0}\frac{A'B'}{AB} = \frac{\mathrm{d}s'}{\mathrm{d}s}. \tag{1.71}$$

According to formulae (1.40, 1.44), Eq. (1.71) can be written as follows:

$$\mu^2 = \frac{\left(e\mathrm{d}q^2 + 2f\mathrm{d}q\mathrm{d}\lambda + g\mathrm{d}\lambda^2\right)\cos^2\alpha}{r^2\mathrm{d}q^2}. \tag{1.72}$$

Let us transform formula (1.72):

$$\mu^2 = \frac{e}{r^2}\cos^2\alpha + \frac{2f}{r^2}\left(\frac{\mathrm{d}\lambda}{\mathrm{d}q}\right)\cos^2\alpha + \frac{g}{r^2}\left(\frac{\mathrm{d}\lambda}{\mathrm{d}q}\right)^2\cos^2\alpha. \tag{1.73}$$

Using Eq. (1.19), we write:

$$\mu^2 = \frac{1}{r^2}\left(e\cos^2\alpha + f\sin 2\alpha + g\sin^2\alpha\right). \tag{1.74}$$

Formula (1.74) is called the **general formula of the linear scale**. In this formula, all quantities are variables. The functions e, g, f and r depend on the coordinates q, λ of a point on the surface of the ellipsoid. This means that the local linear scale μ depends on the coordinates of the point on the surface of the ellipsoid. However, according to formula (1.74), the scale also depends on one more quantity: on the azimuth of the direction, i.e., on α. Consequently, at one point of projection, the scale in different directions is different.

1.6.2 Scale Along the Meridians and Parallels of the Cartographic Grid

Obviously, for scales along meridians and parallels, we can write:

$$m = \mu|_{\alpha=0}, n = \mu|_{\alpha=90}, \tag{1.75}$$

or, with Eq. (1.74),

$$m = \frac{\sqrt{e}}{r}, \tag{1.76}$$

$$n = \frac{\sqrt{g}}{r}. \tag{1.77}$$

Using formulae (1.34, 1.76–1.77), we rewrite Eq. (1.74) in the following form:

$$\mu^2 = m^2\cos^2\alpha + mn\cos i\sin 2\alpha + n^2\sin^2\alpha. \tag{1.78}$$

Formula (1.78) is called the **general formula of the linear scale** (Bugayevskiy and Snyder 1995, p. 18), expressed in terms of linear scales along meridians and parallels.

1.6.3 Main Directions at the Projection Point

As shown earlier, at one point, there are an infinite number of local linear scales, depending on the direction. Among these scales, there are a maximum scale and a minimum scale.

The main directions are the directions at a given projection point along which extremal scales act.

To find the azimuths of the main directions α_0, we must differentiate formula (1.74) with respect to α and equate the resulting derivative to zero.

$$\frac{d\mu^2}{d\alpha}|_{\alpha=\alpha_0} = \frac{1}{r^2}(-e \sin 2\alpha_0 + 2f \cos 2\alpha_0 + g \sin 2\alpha_0) = 0. \tag{1.79}$$

Dividing all terms of Eq. (1.79) by $\cos 2\alpha_0$, we obtain:

$$\tan 2\alpha_0 = \frac{2f}{e-g}. \tag{1.80}$$

Equation (1.80) has two solutions in the interval from $(-\pi)$ to $(+\pi)$. If the first solution is designated $2\alpha_0$, then the second one will differ from the first by π, i.e., $2\alpha_0 \pm \pi$. Thus, Eq. (1.80) has two solutions:

$$\alpha_0, \alpha_0 \pm \frac{\pi}{2}.$$

Therefore, there are two main directions at each point of the projection, and they are orthogonal to each other on the ellipsoid.

1.6.4 Tissot's Indicatrix

At point A on the surface of the ellipsoid, we construct an infinitesimal circle of radius ρ'. Let us choose an arbitrary point B on the circle and an arbitrary Cartesian coordinate system ξ, η on the ellipsoid with a center at point A. Obviously, in this system, the coordinates of point B are equal to ξ, η and satisfy the equation of the circle (Bugayevskiy and Snyder 1995, p. 24; Papadopoulos 2017):

$$\xi^2 + \eta^2 = \rho'^2. \tag{1.81}$$

In the projection onto the plane, the points A and B are represented by the points A' and B'. The axes ξ, η are represented by axes ξ', η'. In the projection, the perpendicularity of the axes may be distorted. Point B' in an oblique coordinate system has coordinates ξ', η'. Obviously, the point B' is on a closed curve, which is an image of a circle (Fig. 1.7).

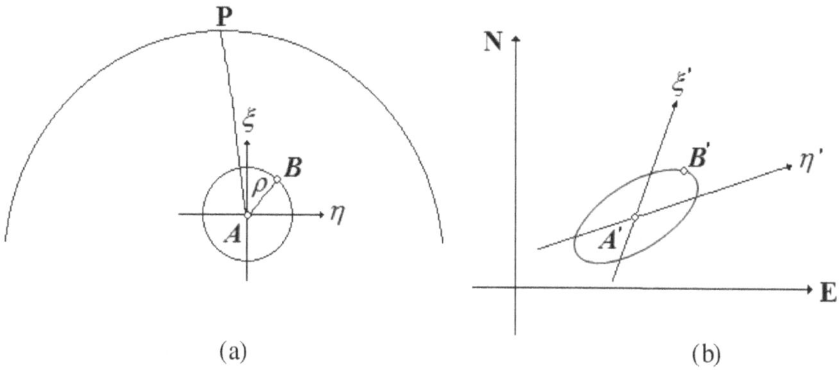

Fig. 1.7 Procedure for constructing Tissot's indicatrix: **a** an elementary circle on the surface of an ellipsoid, and **b** the image of this circle on the plane is Tissot's indicatrix in a curvilinear coordinate system on the plane

Let the scale μ_1 act along the axis ξ and the scale μ_2 act along the axis η. Then:

$$\xi' = \xi \cdot \mu_1, \eta' = \eta \cdot \mu_2. \tag{1.82}$$

Let us find ξ, η from equalities (1.82) and substitute their values into formula (1.81):

$$\frac{\xi'^2}{(\mu_1 \rho')^2} + \frac{\eta'^2}{(\mu_2 \rho')^2} = 1. \tag{1.83}$$

An analysis of formula (1.83) shows that the curve with point B' is elliptical. This curve is called the **indicatrix of distortions**, or after the scientist proposed taking into account linear distortions using this curve, **Tissot's indicatrix** (Tissot 1881).

Thus, **Tissot's indicatrix** is the image on the plane of an infinitesimal circle constructed on the surface of an ellipsoid.

Let us choose a system of rectangular coordinates ξ_0, η_0 on the surface of the ellipsoid that coincides at point A with the main directions. Let there be a maximum scale a along the direction ξ_0 and a minimum scale b along the direction η_0. Since a circle of an arbitrary radius was chosen on the surface of the ellipsoid, we equated its radius to 1. Then, Tissot's indicatrix equation is rewritten as follows:

$$\frac{\xi'^2_0}{a^2} + \frac{\eta'^2_0}{b^2} = 1. \tag{1.84}$$

Notably, the image of the main directions in the plane is also mutually perpendicular, i.e., the axes ξ_0' and η_0' are orthogonal (Bugayevskiy and Snyder 1995, p. 24). Figure 1.8 shows an image of an infinitesimal circle in a coordinate system related to the main directions.

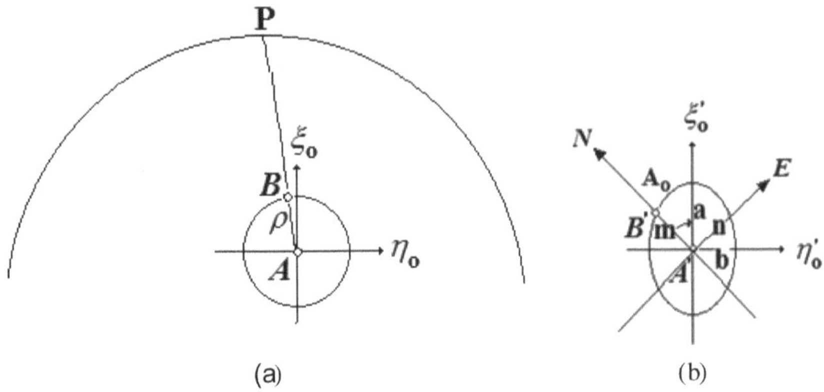

Fig. 1.8 Construction of Tissot's indicatrix in a coordinate system coinciding with the main directions; **a** a circle of an infinitesimal radius, constructed in a coordinate system coinciding with the main directions on the surface of the ellipsoid, and **b** the corresponding Tissot's indicatrix on the plane

To find extremal scales, we use two theorems of Apollonius for the semidiameters of an ellipse (Bugayevskiy and Snyder 1995, p. 19):

1. The sum of the squares of the main semidiameters is equal to the sum of the squares of the conjugate semidiameters.
2. The product of the main semidiameters is equal to the product of the conjugate semidiameters multiplied by the sine of the angle between them.

For Tissot's indicatrix, the main semidiameters are the extremal scales a, b; the conjugate semidiameters are the scales along the meridians and parallels m, n; and the angle between the conjugate semidiameters is the angle between the meridian and the parallel of the cartographic grid, equal to i. Thus, Apollonius' theorems can be written as follows:

$$a^2 + b^2 = m^2 + n^2, ab = mn \sin i, \tag{1.85}$$

or taking into account (1.34) and (1.76–1.77),

$$a^2 + b^2 = \frac{e + g}{r^2}, ab = \frac{h}{r^2}. \tag{1.86}$$

Let us multiply the second Eq. (1.86) by 2 and sum it with the first equation. We obtain:

$$(a + b)^2 = \frac{e + g + 2h}{r^2}. \tag{1.87}$$

Similarly, subtracting the doubled second equation from the first equation of Apollonius's theorems, we have:

$$(a-b)^2 = \frac{e+g-2h}{r^2}. \tag{1.88}$$

Let us find the extremal scales from equalities (1.87) and (1.88):

$$a = \frac{\widetilde{A}+\widetilde{B}}{2}, b = \frac{\widetilde{A}-\widetilde{B}}{2}, \tag{1.89}$$

where

$$\widetilde{A} = \frac{1}{r}\sqrt{e+g+2h}, \widetilde{B} = \frac{1}{r}\sqrt{e+g-2h}. \tag{1.90}$$

1.6.5 Orientation of the Indicatrix on the Plane

To plot Tissot's indicatrix on a map, it is necessary to know not only the values of extremal scales a, b but also the grid azimuth of the direction of the maximum scale $_0$ (see Fig. 1.8b).

Let us find a formula for determining the grid azimuth on a plane, using the parameters of the inverse projection rather than the direct one.

The projection from a plane onto the surface of a sphere or ellipsoid, because point A' is mapped to the original point A (see Fig. 1.1), is called the **inverse projection** of the original map projection.

Inverse projection formulae in general can be written as follows (Bugayevskiy and Snyder 1995, p. 13):

$$q = f_1'(N,E), \ = f_2'(N,E). \tag{1.91}$$

According to the definition of the forward projection length scale, the inverse projection scale is equal to:

$$\mu' = \lim_{A'B' \to 0} \frac{AB}{A'B'} = \frac{\mathrm{d}s}{\mathrm{d}s'} = \frac{1}{\mu}. \tag{1.92}$$

The total differentials of the inverse projection mapping functions, by analogy with equalities (1.16), are written as follows:

$$\mathrm{d}q = q_N \mathrm{d}N + q_E \mathrm{d}E, \ \mathrm{d}\lambda = \lambda_N \mathrm{d}N + \lambda_E \mathrm{d}E. \tag{1.93}$$

Substituting (1.93) into (1.39) yields:

$$\mathrm{d}s^2 = r^2\left(e'\mathrm{d}N^2 + 2f'\mathrm{d}N\mathrm{d}E + g'\mathrm{d}E^2\right), \tag{1.94}$$

where

$$e' = q_N^2 + \lambda_N^2, g' = q_E^2 + \lambda_E^2, f' = q_N q_E + \lambda_N \lambda_E. \tag{1.95}$$

Similar to formula (1.40), the linear element on the plane is equal to:

$$\mathrm{d}s'^2 = \mathrm{d}N^2 + \mathrm{d}E^2 = \mathrm{d}N^2\left(1 + \frac{\mathrm{d}E^2}{\mathrm{d}N^2}\right) = \mathrm{d}N^2\left(1 + \tan^2 A\right) = \frac{\mathrm{d}N^2}{\cos^2 A}. \tag{1.96}$$

Taking into account (1.92, 1.94, 1.96), we obtain:

$$\mu'^2 = r^2\left(e' \cos^2 A + f' \sin 2A + g' \sin^2 A\right), \tag{1.97}$$

where A is the grid azimuth of the direction along which the scale μ' acts.

To determine the grid azimuth of the main direction, for the direct projection, we find the derivative of the function μ' and equate it to zero. After identical transformations, we get:

$$\left.\frac{\partial \mu'^2}{\partial A}\right|_{A=A_0} = r^2\left(-e' \sin 2A_0 + 2f' \cos 2A_0 + g' \sin 2A_0\right) = 0, \tag{1.98}$$

or,

$$\tan 2A_0 = \frac{2f'}{e' - g'}. \tag{1.99}$$

Formula (1.99) is completely identical to formula (1.80) for calculating the azimuth of the main direction on the original surface. However, it determines the grid azimuths of the main directions in the plane, not for the forward projection but for the inverse projection. If we take into account the formula for connecting the scales of forward and inverse projections (1.92), then we can say that the grid azimuths of the main directions of forward and inverse projections coincide. The only difference is that the direction of the maximum scale of the forward projection is the direction of the minimum scale of the inverse projection and vice versa. All the abovementioned methods apply not only for grid azimuths on the plane but also for azimuths on the original surface. They are also the same for forward and inverse projections.

Applying the hodograph transformation (Bugayevskiy and Snyder 1995, p. 173; Rozhdestvensky and Yanenko 1978, p. 33):

$$q_N = E_\lambda \frac{1}{h}, q_E = -N_\lambda \frac{1}{h}, \lambda_N = -E_q \frac{1}{h}, \lambda_E = N_q \frac{1}{h}, \tag{1.100}$$

let us move from the mapping functions of the inverse projection in formula (1.99) to the mapping functions of the forward projection. We obtain:

$$\tan 2A_0 = \frac{2\left(N_q E_q + N_\lambda E_\lambda\right)}{\left(N_q^2 + N_\lambda^2 - E_q^2 - E_\lambda^2\right)}. \tag{1.101}$$

As shown below, the azimuths of the main directions on the original surface and the grid azimuths of the main directions on the plane play important roles in posing and solving the problem of finding the best projections.

1.7 Local Area Scale

Let us select a point A with coordinates q, λ on the surface of the ellipsoid and a point C infinitely close to it with coordinates $q + \mathrm{d}q$, $\lambda + \mathrm{d}\lambda$. Let us draw meridians and parallels through these points. They create an elementary spheroidal trapezoid on the surface of the ellipsoid (Fig. 1.9a). The area of this trapezoid is as follows:

$$\mathrm{d}\sigma = r^2 \mathrm{d}q\mathrm{d}\lambda. \tag{1.102}$$

An elementary spheroidal trapezoid ABCD on the plane is represented by an elementary parallelogram $A'B'C'D'$. The sides of this parallelogram are the elements of the arcs of meridians and parallels, namely:

$A'B'$ is an element of the meridian arc on the plane;

$A'D'$ is an element of the parallel arc on the plane.

These arcs can be determined by using formula (1.44):

$$\mathrm{d}s'_{A'B'} = \mathrm{d}s'\big|_{d\lambda=0} = \sqrt{e}\mathrm{d}q, \; \mathrm{d}s'_{A'D'} = \mathrm{d}s'\big|_{\mathrm{d}q=0} = \sqrt{g}\mathrm{d}\lambda. \tag{1.103}$$

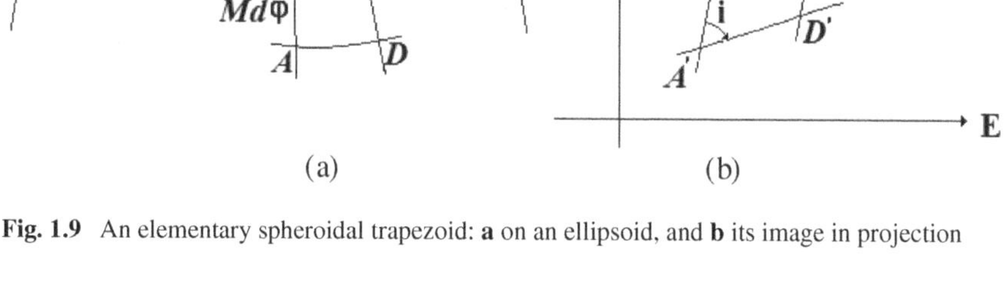

Fig. 1.9 An elementary spheroidal trapezoid: **a** on an ellipsoid, and **b** its image in projection

The area of the elementary parallelogram is equal to:

$$d\sigma' = ds'_{A'B'} \cdot ds'_{A'D'} \cdot \sin i = \sqrt{eg} \sin i dq d\lambda. \tag{1.104}$$

According to (Bugayevskiy and Snyder 1995, p. 20), the **local area scale** is the limit of the ratio of the area of an elementary figure in projection to the area of the corresponding figure on the ellipsoid when the area of the latter tends to zero, namely:

$$p = \lim_{d\sigma \to 0} \frac{d\sigma'}{d\sigma} = \frac{\sqrt{\text{eg}} \sin i dq d\lambda}{r^2 dq d\lambda}. \tag{1.105}$$

Considering the formulae for linear scales along meridians and parallels (1.76–1.77), formula (1.105) can be rewritten as follows:

$$p = mn \sin i. \tag{1.106}$$

If we use Apollonius's second theorem (equality (1.85)), then formula (1.106) can be written as follows:

$$p = ab. \tag{1.107}$$

Another equation for the area scale can be obtained by replacing the function sin i in Eq. (1.105) with its value according to the first formula of the system (1.34):

$$p = \frac{|h|}{r^2}. \tag{1.108}$$

In classical mathematical cartography, it is a priori assumed that both the linear and area scales are nonnegative. Moreover, the formulae for these quantities do not indicate the determination of the sign of these quantities, except for formula (1.108), which contains a Jacobian in its numerator. The sign of the Jacobian depends on the choice of coordinate systems both on the surface and on the plane, as well as on the choice of the design procedure for positive and negative projections (see Fig. 1.3). For the area scale to be nonnegative everywhere, the absolute value of the Jacobian must be in the numerator of formula (1.108).

As discussed below, in the theory of best projections, the change in the sign of the Jacobian cannot be ignored since this change indicates whether the projection satisfies the one-to-one principle. In this regard, to indicate mutual unambiguity, an indicator is equal to:

$$\overline{\sigma} = \frac{h}{R^2}. \tag{1.109}$$

Unlike the linear and area scales, which are dimensionless, the Jacobian has a dimension corresponding to the dimension of the square of the coordinates of the points. To make the indicator dimensionless, a value equal to the square of the radius of the sphere is inserted into the denominator of the formula for its calculation.

For the study of the mutual uniqueness of the projection, the size of the sphere does not matter. In this way, we can establish:

$$R = 1.$$

Then:

$$\overline{\sigma} = \frac{h}{R^2} = \frac{h}{1} = h. \quad (1.110)$$

The magnitude and sign of the Jacobian are not invariant with respect to the chosen coordinate system. Only the change in sign is invariant. Therefore, by the sign of the Jacobian, it is impossible to determine which part of the projection is positive and which part is negative (see Fig. 1.3). At the same time, it is possible to select a coordinate system for which the positive part of the projection corresponds to a positive value of the Jacobian, and the negative part of the projection corresponds to a negative value of the Jacobian. This coordinate system is called **positively oriented**.

1.8 Types of Distortions in Projection

The main projection distortions are distortions of angles, line lengths, and areas. Let us look at these distortions in more detail.

1.8.1 Distortion of Angles

Let us construct an infinitesimal circle on the surface of the ellipsoid at a point A and choose a rectangular coordinate system ξ_0, η_0 coinciding with the main directions at this point. Let us choose point B of the circle. The direction AB is measured by the angle u on an ellipsoid (Fig. 1.10):

$$\tan u = \frac{\eta_0}{\xi_0}. \quad (1.111)$$

On a plane, point A is represented by point A', point B is represented by point B', a circle is represented by an ellipse, called Tissot's indicatrix, and a direction u is represented by a direction u'. On the plane, the direction u' is determined by the following formula:

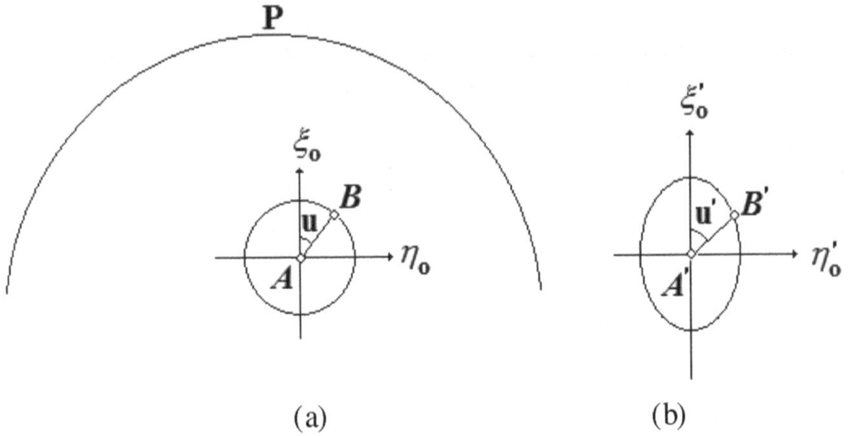

Fig. 1.10 Distortion of direction: **a** the original direction on the ellipsoid, and **b** its distorted image

$$\tan u' = \frac{\eta'_0}{\xi'_0}. \tag{1.112}$$

Since the maximum scale acts along the direction ξ_0 and the minimum scale acts along the direction η_0, then:

$$\xi'_0 = \xi_0 \cdot a, \eta'_0 = \eta_0 \cdot b. \tag{1.113}$$

Substituting (1.113) into (1.111) yields:

$$\tan u' = \frac{b}{a} \tan u. \tag{1.114}$$

Let us write the obvious equality:

$$\tan u = \tan u. \tag{1.115}$$

Equalities (1.114) and (1.115) are added and subtracted, resulting in the following:

$$\begin{array}{l} \tan u + \tan u' = \left(1 + \frac{b}{a}\right) \tan u \\ \tan u - \tan u' = \left(1 - \frac{b}{a}\right) \tan u \end{array}. \tag{1.116}$$

Having performed identical transformations of the system (1.116), we obtain:

$$\begin{array}{l} \frac{\sin(u+u')}{\cos u \cos u'} = \frac{a+b}{a} \tan u \\ \frac{\sin(u-u')}{\cos u \cos u'} = \frac{a-b}{a} \tan u \end{array}. \tag{1.117}$$

Let us divide the second equation of the system (1.117) by the first equation:

$$\sin(u - u') = \sin(u + u')\frac{a - b}{a + b}. \tag{1.118}$$

According to Eq. (1.118), the distortion of direction $(u - u')$ depends on the direction value u itself. The main directions do not have distortions. Obviously, the most distorted direction u_0 satisfies the following condition:

$$\sin(u_0 + u'_0) = 1. \tag{1.119}$$

In formula (1.119), u'_0 is the image on the plane of the most distorted direction. Solving Eq. (1.119), we obtain:

$$u_0 + u'_0 = \frac{\pi}{2}. \tag{1.120}$$

Let us denote the distortion of the most distorted direction by $\frac{\omega}{2}$, i.e.,

$$u_0 - u'_0 = \frac{\omega}{2}. \tag{1.121}$$

Using formulae (1.118–1.121), we obtain:

$$\sin\frac{\omega}{2} = \frac{a - b}{a + b}. \tag{1.122}$$

If an angle consists of the two most distorted directions, then its distortion is equal to ω.

Using trigonometry formulae, we obtain (Dugayevskiy and Snyder 1995, p. 24):

$$\cos\frac{\omega}{2} = \sqrt{1 - \sin^2\frac{\omega}{2}} = 2\frac{\sqrt{ab}}{a + b}, \quad \tan\frac{\omega}{2} = \frac{a - b}{2\sqrt{ab}}, \tag{1.123}$$

$$\tan\frac{\omega}{4} = \frac{\sqrt{a} - \sqrt{b}}{\sqrt{a} + \sqrt{b}}, \quad \tan\left(\frac{\pi}{4} + \frac{\omega}{4}\right) = \sqrt{\frac{a}{b}}. \tag{1.124}$$

Equality (1.122) allows us to estimate the maximum amount of angular distortion. To determine the real distortion of angles in specific fixed directions, let us return to formula (1.118). The azimuth of the direction u on the ellipsoid is equal to α. Since the azimuth of the main direction from which the angle u is measured is equal to α_0, the magnitude of the angle u is as follows:

$$u = \alpha - \alpha_0. \tag{1.125}$$

Similarly, the magnitude of the angle u' is:

$$u' = \alpha' - \alpha'_0. \tag{1.126}$$

The value $(u - u')$ is determined by the following relation:

$$\sin(u - u') = \sin(\alpha - \alpha_0 + \alpha' - \alpha'_0)\frac{a - b}{a + b}. \tag{1.127}$$

Taking into account (1.125–1.126), we determine the relationship between azimuth distortion and distortion in the direction u:

$$u - u' = \alpha - \alpha' - (\alpha_0 - \alpha'_0). \tag{1.128}$$

Hence, (Bugayevskiy and Snyder 1995, p. 25)

$$\alpha - \alpha' = (u - u') + (\alpha_0 - \alpha'_0). \tag{1.129}$$

Formula (1.129) is a formula for the relationship between angular distortion $(u - u')$ and azimuth distortion $(\alpha - \alpha')$.

1.8.2 Distortion of Length

Let us divide the procedure for mapping the surface of an ellipsoid onto a plane into two parts. In the first stage, the surface of the ellipsoid is mapped onto the surface of an ellipsoidal globe, similar to the Earth's ellipsoid. In the second stage, the surface of the globe is unfolded onto a plane (Fig. 1.11).

Let us denote:

$$\mu_0 = \frac{ds_0}{ds} = \text{const}, \, c = \frac{ds'}{ds_0}. \tag{1.130}$$

In formulae (1.130), μ_0 is the **nominal linear scale**; c is **the scale distortion**. As indicated in (Bugayevskiy and Snyder 1995, p. 17), the nominal scale is a constant value.

Using the formula for the local length scale (1.71) and equality (1.130), we write the formula for the connection between the local scale and the nominal scale:

$$\mu = c\mu_0. \tag{1.131}$$

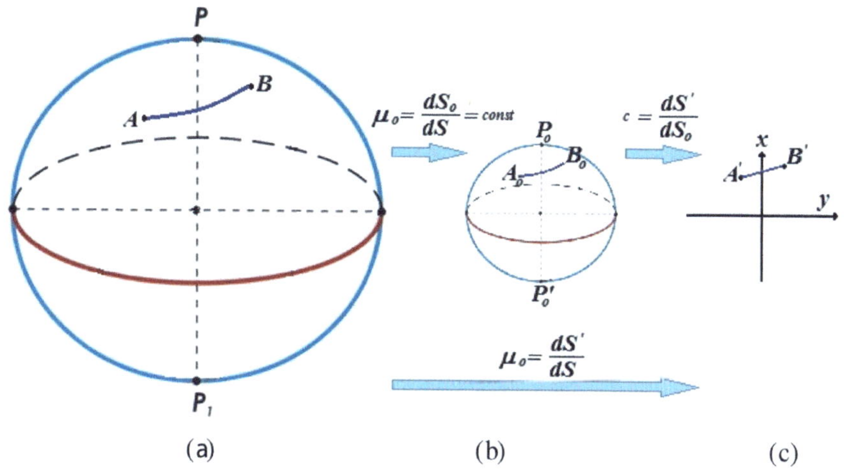

Fig. 1.11 Procedure for mapping the surface of an ellipsoid onto a plane, divided into stages: **a** geodesic line on the surface of the Earth's ellipsoid, **b** geodesic line on the surface of a similar ellipsoidal globe, and **c** straight line on a plane

Let us write formula (1.131) in the following form:

$$c = \frac{\mu}{\mu_0}. \quad (1.132)$$

The scale distortion is a number that determines the quality of linear distortion in the projection. The closer this number is to one, the better the projection is considered. The first measure of linear distortion is the magnitude of the deviation of the scale distortion from 1, namely:

$$V = c - 1, \quad (1.133)$$

where V is the scale error or linear distortion (Snyder 1987, p. 21).

When studying the properties of a projection, the nominal scale is not taken into account. Therefore, in formula (1.132) μ_0 is equated to unity. In this case, formula (1.133) takes the form:

$$V = \mu - 1. \quad (1.134)$$

Since, in this case, μ is not an absolute but a relative value, it is called a **linear scale factor**. Similarly, the relative magnitudes of the extremal scales of the lengths are called the maximum and minimum scale factors, respectively. The corresponding maximum V_a and minimum V_b linear distortions are equal to:

$$V_a = a - 1, V_b = b - 1. \quad (1.135)$$

Traditionally, the linear distortions are determined as a percentage, i.e., multiplied by 100.

For more information on determining linear distortions in projections, see Sect. 2.2.

1.8.3 Area Distortion

Area distortion in projection is the deviation of the area scale factor from unity, i.e.,

$$V_p = p - 1. \tag{1.136}$$

As in the previous case, this value is determined as a percentage.

1.9 Classification of Projections by Distortion Characteristics

This classification is based on equalities that are satisfied by extremal scale factors. According to this classification, projections are divided into four groups: conformal, equal-area, equidistant, and arbitrary in the nature of distortions. Let us look at them in more detail.

1.9.1 Conformal Projections

In these projections, the angles are not distorted:

$$\omega = 0. \tag{1.137}$$

Taking into account formula (1.122), we write for conformal projections:

$$a = b. \tag{1.138}$$

Thus, in conformal projections, the extremal linear scale factors are equal. This means that the linear scale factor does not depend on direction. In any direction, the scale factor has the same value. Tissot's indicatrixes in conformal projections are circles. Therefore, we can conclude that in conformal projections, infinitesimal circles are displayed as circles.

1.9.2 Equal-Area Projections

The equal-area condition is written as follows:

$$p = 1, \tag{1.139}$$

or, taking into account formula (1.107):

$$ab = 1. \tag{1.140}$$

Tissot's indicatrixes in equal-area projections are ellipses. However, the areas of these ellipses are always the same.

1.9.3 Equidistant Projections

There are no projections that are completely free of line length distortions. However, there are projections in which there are no distortions along one of the main directions. These projections are called equidistant. The equidistant condition is written as follows:

$$a = 1, \text{ or } b = 1. \tag{1.141}$$

If the equidistant projection is also orthogonal, then taking into account formulae (1.85), we can write:

$$a = \max(m, n)\ b = \min(m, n) \tag{1.142}$$

If

$$a = m = 1, \text{ or } b = m = 1, \tag{1.143}$$

then the projection is called equidistant along the meridians.

Likewise, if

$$a = n = 1, \text{ or } b = n = 1, \tag{1.144}$$

the projection is called equidistant along the parallels.

If one of the extremal scale factors in the mapping area is equal not to unity but to some constant value, then the projection is called equidistant in the broad sense.

1.9.4 Projections with Arbitrary Distortions

If the extremal scale factors of the projection do not satisfy any of the conditions (1.138, 1.140, 1.141), the projection is considered arbitrary in terms of the nature of the distortions.

1.10 Comparison of Distortions in Projections Determined by the Distortion Characteristics

Let us assume that three projections are constructed on one area: conformal, equal-area, and equidistant, and that the linear distortions in all three projections are quantities of the first order of smallness.[1]

As shown by Kavraysky (1933, 1934), if in the projection, the linear distortions are quantities of the first order of smallness, then the distortions of areas and angles are also quantities of the first order of smallness. Indeed, according to (Bugayevskiy and Snyder 1995, p. 28), taking into account equalities (1.135), the area distortions are equal to:

$$V_p = (ab - 1) = (V_a + 1)(V_b + 1) - 1 = V_a \cdot V_b + V_a + V_b. \tag{1.145}$$

If linear distortions V_a, V_b are quantities of the first order of smallness, then $V_a \cdot V_b$ are quantities of the second order of smallness. Therefore, up to quantities of the second order of smallness:

$$V_p \approx V_a + V_b. \tag{1.146}$$

Returning to extremal scale factors using equalities (1.135), we obtain:

$$V_p \approx a + b - 2. \tag{1.147}$$

For angle distortions, according to (1.122), we can write:

$$\omega = 2\arcsin\left(\frac{a - b}{a + b}\right). \tag{1.148}$$

[1] The order of smallness is defined by infinitesimal quantities (Bronstejn et al., 2020). In our case, small but finite quantities occur. The question arises: what quantity can be considered small of the first order? The answer to this question can be found in the analysis of the sum of the geometric progression.

Let us consider the sum of a geometric progression $x + x^2 + x^3 + \ldots$ as the sum of the quantities of the first order of smallness (x), the second order of smallness (x^2), etc. As is known (Bronstejn et al., 2020), the sum of a geometric progression exists only in the case when $|x| < 1$. Obviously, a quantity of the first order of smallness must satisfy this inequality.

Substituting equalities (1.135) into (1.148) yields:

$$\omega = 2\arcsin\left(\frac{V_a - V_b}{2 + V_a + V_b}\right). \tag{1.149}$$

Since linear distortions are first-order quantities of smallness, angular distortions are also first-order quantities of smallness. Therefore, instead of the arcsine function, we can write its argument in formula (1.149), i.e.,

$$\omega \approx 2\left(\frac{V_a - V_b}{2 + V_a + V_b}\right). \tag{1.150}$$

The denominator of the formula (1.150) contains a value close to 2. Therefore, we write:

$$\omega \approx V_a - V_b. \tag{1.151}$$

Formulae (1.146) and (1.151) confirm the statement that if the linear distortions of the projection are of the first order of smallness, then the distortions of the angles and areas are also of the first order of smallness.

Let us study the behavior of projection distortions, which is determined by the distortion characteristics.

1.10.1 Conformal Projections

Taking into account equalities (1.135) and (1.138), we can write:

$$V = V_a = V_b. \tag{1.152}$$

According to formula (1.146), we write for the area distortions of conformal projections:

$$V_p \approx V_a + V_b = 2V. \tag{1.153}$$

1.10.2 Equidistant Projections

Specifically, let us assume that in the equidistant projections, the minimum scale factor is equal to unity, i.e.,

$$b = 1, V_b = 0. \tag{1.154}$$

Then, the extremal linear distortions in these projections are equal to:

$$V = V_a = a - 1. \tag{1.155}$$

Let us find the area distortions of the equidistant projections. Taking into account (1.154), formula (1.146) takes the following form:

$$V_p \approx V_a + V_b = V_a = V. \tag{1.156}$$

Using formulae (1.151) and (1.155), we write:

$$\omega \approx V_a - V_b = V_a = V. \tag{1.157}$$

Thus, in equidistant projections, the distortions of angles, line lengths, and areas are almost equal.

1.10.3 Equal-Area Projections

Taking into account (1.146) and the fact that area distortions in equal-area projections are equal to zero, we obtain:

$$V_a \approx -V_b. \tag{1.158}$$

Consequently, in equal-area projections, extremal linear distortions, up to values of the second order of smallness, are equal in absolute value but have different signs. Taking into account the formulae (1.151, 1.158), we write:

$$\omega \approx V_a - V_b = 2V. \tag{1.159}$$

All the abovementioned points are reflected in the Table 1.1.

According to Table 1.1, all distortions of equidistant projections are balanced and approximately equal to each other. In contrast, in conformal projections, the area distortions are twice as large as the linear distortions. In equal-area projections, the

Table 1.1 Comparison of distortions in projections determined by the distortion characteristics

Projection	Distortions of		
	Angles	Line lengths	Areas
Conformal	0	V	$\approx 2V$
Equidistant	$\approx V$	V	V
Equal-area	$\approx 2V$	V	0

situation is reversed: the distortions of angles are twice as large as the distortions of line lengths.

Table 1.1 clearly demonstrates the principle of the arithmetic mean in projections (see Sect. 4.5 for more details). According to this principle, if two projections, conformal and equal-area, are known for the same mapping area and if the linear distortions in these projections are of the same order, then the arithmetic mean of the mapping functions of both projections will yield a third projection. In this third projection, the angular distortions are half those of the original equal-area projection, and the area distortions are half those of the original conformal projection.

Notably, the first of the ideal projections known to humanity, the Postel projection in the direct version, is a projection equidistant along the meridians.

1.11 Coefficients of Metric Shape of a Surface

The metric surface shape coefficient is included in all the formulae for linear scale factors (1.74, 1.76, and 1.77) and area scale factors (1.108). Therefore, it plays an important role in the theory of the best projections. As mentioned above, if the coordinate system q, λ is used on the surface of the ellipsoid, the coefficient of the metric shape of the surface, r, is the radius of the parallel (see formula (1.39)). Thus, the surface metric shape coefficient is a characteristic of the surface properties. However, the surface metric shape coefficient is not invariant with respect to the chosen coordinate system. The transition from one coordinate system to another leads to a change in this function. Let us choose another isometric coordinate system q', λ' different from q, λ. Since both systems are isometric, they contain conjugate harmonic functions, and their first derivatives satisfy the Cauchy–Riemann system of equations, namely (Bugayevskiy and Snyder 1995, p. 31):

$$q_{q'} = \pm\lambda_{\lambda'}, \quad q_{\lambda'} = \mp\lambda_{q'}. \tag{1.160}$$

Let us write the formulae of total differentials for the coordinates q, λ with respect to the variables q', λ'.

$$\mathrm{d}q = q_{q'}\mathrm{d}q' + q_{\lambda'}\mathrm{d}\lambda', \quad \mathrm{d}\lambda = \lambda_{q'}\mathrm{d}q' + \lambda_{\lambda'}\mathrm{d}\lambda'. \tag{1.161}$$

Substituting (1.161) into (1.39) leads to:

$$\mathrm{d}s^2 = r^2\left[\left(q_{q'}^2 + {}_{q'}^2\right)\mathrm{d}q'^2 + 2\left(q_{q'}q_{\lambda'} + {}_{\lambda q'}\lambda_{\lambda'}\right)\mathrm{d}q'\mathrm{d}\lambda' + \left(q_{\lambda'}^2 + \lambda_{\lambda'}^2\right)\mathrm{d}\lambda'^2\right], \tag{1.162}$$

or taking into account (1.160):

$$\mathrm{d}s^2 = r^2\left(q_{q'}^2 + q_{\lambda'}^2\right)\left(\mathrm{d}q'^2 + \mathrm{d}\lambda'^2\right). \tag{1.163}$$

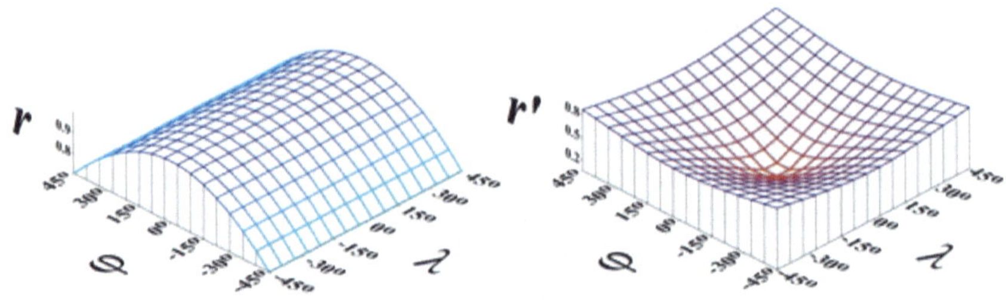

(a) (b)

Fig. 1.12 Three-dimensional graphs of two variants of the coefficient of the metric shape of the surface of the sphere, constructed for the isometric coordinate systems: **a** q, λ, where q is the isometric latitude determined from Eq. (1.12), λ is the geodetic longitude, and **b** q', λ', used to construct the Airy projection in the transverse version (see Sect. 1.5)

Let us denote:

$$r' = r\sqrt{q_{q'}^2 + q_{\lambda'}^2}. \tag{1.164}$$

Then, formula (1.163) can be rewritten as follows:

$$ds^2 = r'^2\left(dq'^2 + d\lambda'^2\right). \tag{1.165}$$

Thus, in the new coordinate system q', λ', the coefficient of the metric shape of the surface is a function defined by formula (1.164). This means that the metric shape coefficient of a surface depends not only on the properties of the surface but also on the chosen isometric coordinate system.

Figure 1.12 shows three-dimensional graphs of two variants of the metric surface shape coefficient for the same territory on the surface of the sphere, with a conditional radius equal to one, limited by meridians $\lambda = \pm 45$ and parallels $\varphi = \pm 45$.

The surface metric shape coefficient in Fig. 1.12a is built on the basis of the following formula that determines the radius of a parallel on a sphere with a radius equal to one, namely:

$$r = \cos\varphi. \tag{1.166}$$

Figure 1.12b shows a graph of the surface metric shape coefficient for the case of a transverse azimuthal projection, in which the relationship between the spherical polar coordinates and geodetic coordinates is described by formulae (5.49) (see Sect. 1.5). The function of the metric form coefficient in this case is the almucantar radius, determined by the following formula:

$$r' = \sin z = \sqrt{1 - \cos^2\varphi\cos^2\lambda}. \tag{1.167}$$

1.12 Visual Representation of Distortions on Map Projection Grids

The most commonly used method of showing distortion on map projection grids is through isolines of equal scale factors or equal distortions, called isocols (Bugayevskiy and Snyder 1995, p. 49). These usually depict extremal linear distortions, V_a, V_b. Combining two isoline graphs into one figure is generally impossible. However, if the cartographic grid has symmetry, the isoline graphs can be combined into one composite graph along the line of symmetry, as shown in Fig. 1.13.

Figure 1.13 illustrates the Bonn projection map grid with isocols of maximum linear distortions, V_a (left side), and minimum linear distortions, V_b (right side of the figure).

The Bonn projection is constructed using the formulae presented in Snyder (1987, p. 138), with a standard parallel of 30°. The meridians and parallels of the cartographic grid are drawn every 30°.

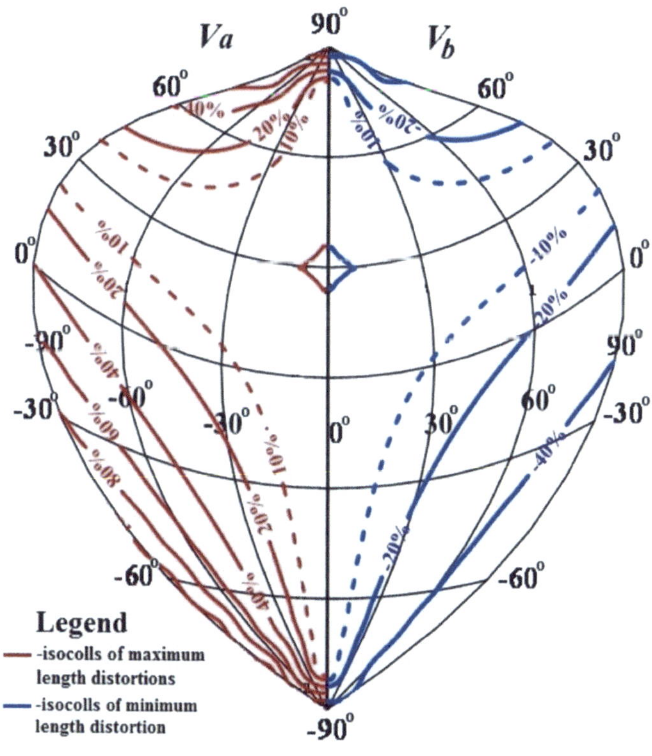

Fig. 1.13 Isocols of maximum linear distortions (left side of the figure) and minimum linear distortions (right side of the figure) for the Bonn projection

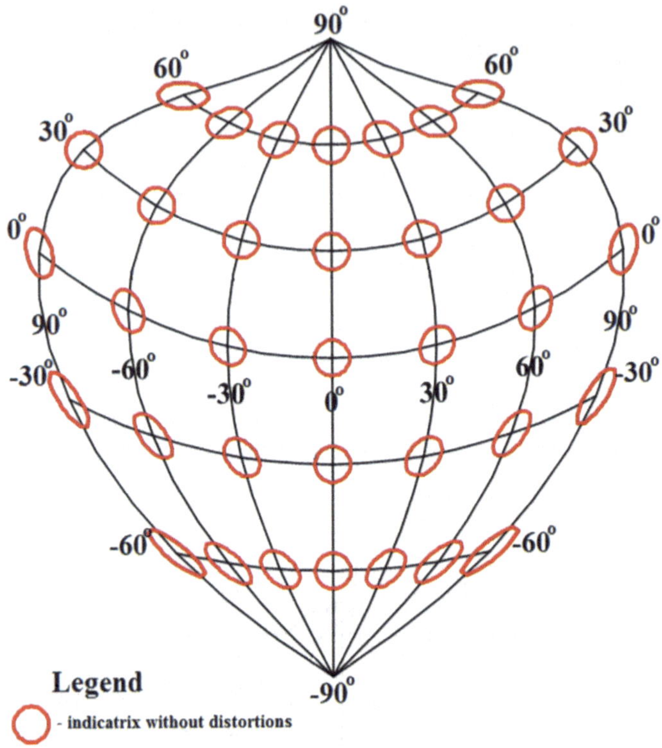

Fig. 1.14 Tissot's indicatrixes on the Bonn projection map grid

Considering that the linear scale factor depends not only on the coordinates of the points but also on the direction at each point of the projection, a more visual way of displaying the linear distortions in the projection is through Tissot's indicatrixes. Figure 1.14 shows Tissot's indicatrixes of the same Bonn projection.

Indicatrixes provide information about the image in the projection of an infinitesimal circle. Serapinas (2005) suggested using the so-called circle of distortions, which in projection shows the image of a finite circle (see Fig. 1.15).

Both Tissot's indicatrixes and distortion circles do not actually show projection distortions. The indicatrix illustrates changes in the linear scale factor at the projection point depending on the direction, whereas the distortion circle shows changes in the final distances at the projection point. Moreover, both indicatrixes and distortion circles are not sensitive to small distortions. Figure 1.16 shows an enlarged fragment of the Bonn projection map grid.

Despite the occurrence of distortions in the projection, they are difficult to assess from Fig. 1.16a, as almost all indicatrixes appear as circles. The human eye cannot discern small changes in the shape of indicatrixes. The symbol-charts are more visual and sensitive to small changes in distortions. They depict distortions rather than scale factors and show linear distortions at a point along a selected direction. Unlike linear

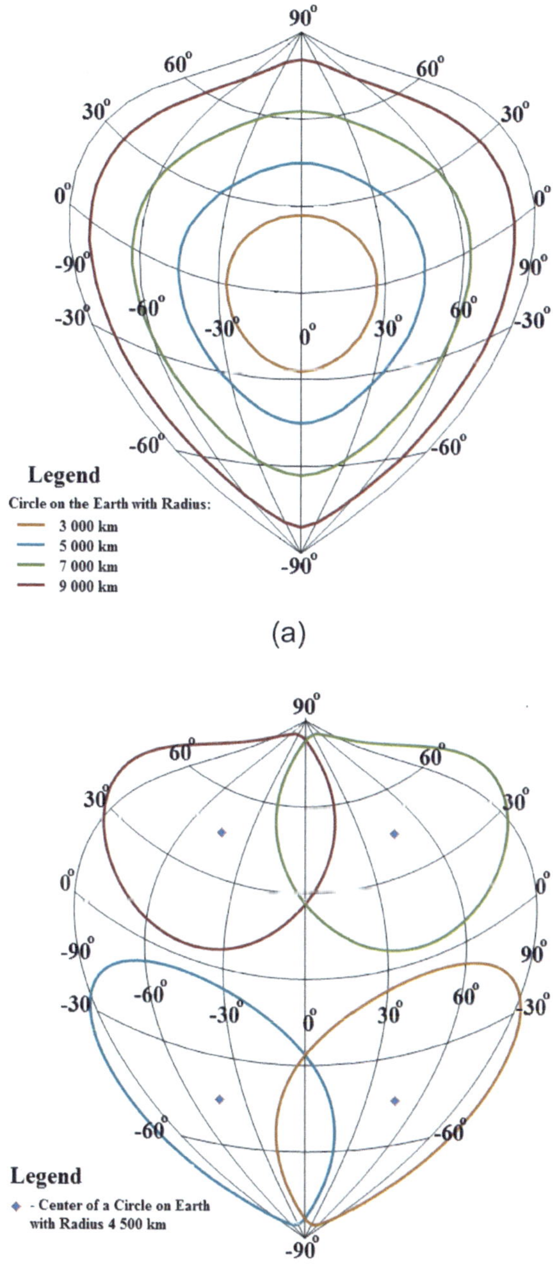

Fig. 1.15 Distortion circles on the Bonn projection: **a** images of concentric circles with a center at a point with coordinates $\varphi = 0$, $\lambda = 0$ and different radii, and **b** images of circles with centers at points with coordinates $\varphi = \pm 45$, $\lambda = \pm 45$ and a radius of 4500 km on the surface of the sphere

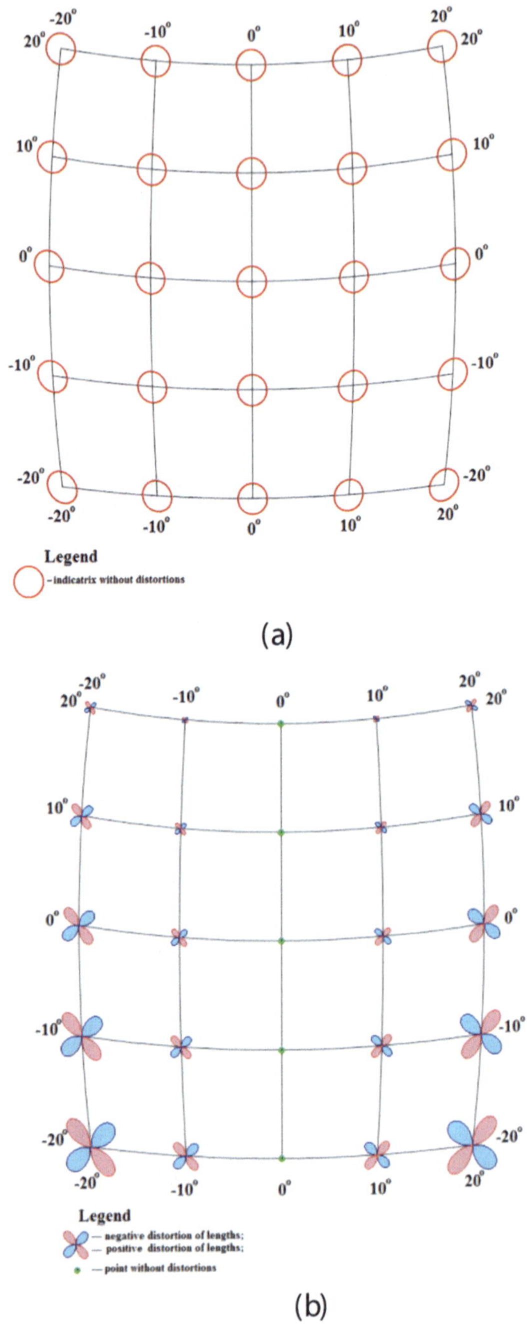

Fig. 1.16 Fragment of the Bonn projection map grid: **a** with Tissot's indicatrixes, and **b** with the symbol-charts at the projection points

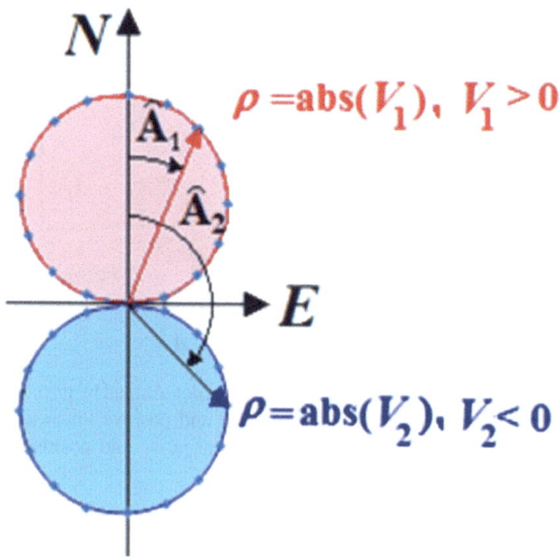

Fig. 1.17 Process of constructing the symbol-chart at a projection point

scale factors, distortion values can be positive or negative. Therefore, the symbol-charts are two-color (Fig. 1.17). Positive distortions are shown with a red curve, whereas negative distortions are shown with a blue curve. In the absence of linear distortion, the curve becomes a point. For clarity, the point in Fig. 1.16b is replaced by a green circle.

The positive and negative distortion parts of the curve are connected by two central points coinciding with the point for which the symbol-chart curve is plotted.

The procedure for constructing a symbol-chart is shown in Fig. 1.17. Symbol-charts are constructed on the same principle as Tissot's indicatrixes. According to Bugayevskiy and Snyder (1995, p. 24), the indicatrix shows the linear scale factor at a certain projection point in a specific direction, whereas the symbol-chart shows the value of the distortion at a certain projection point in a specific direction.

Each symbol-chart is created in a polar coordinate system centered at a grid node. The axis of the polar system coincides with the direction of the N-axis at this point. An angle equal to the grid azimuth is plotted clockwise from this axis:

$$\widehat{A}_k = \arctan \frac{E_q \cos \alpha_k + E \sin \alpha_k}{N_q \cos \alpha_k + N \sin \alpha_k}. \tag{1.168}$$

A distance equal to $|V|$ is plotted in the obtained direction.

Using symbol-charts makes it easy to classify a projection by the nature of its distortions. In particular, for conformal projections, the symbol-charts are circles. For equal-area projections, the symbol-charts have four petals, with equal areas for red and blue petals. For equidistant projections, the symbol-charts have only two lobes, either red or blue (Fig. 1.18).

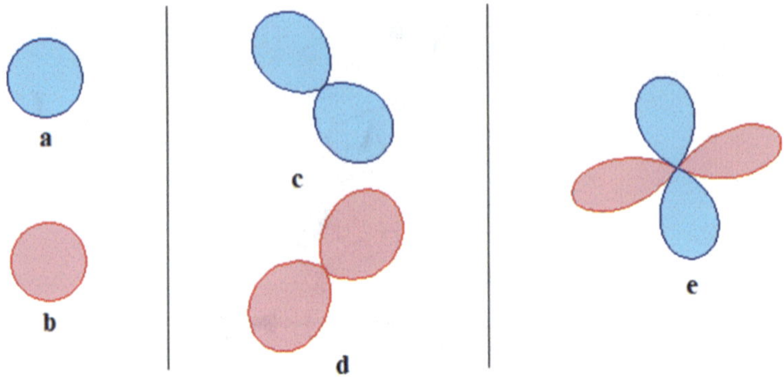

Fig. 1.18 Symbol-charts for projections determined by the nature of distortions: **a**, **b** symbol-charts of conformal projection with negative and positive linear distortions, respectively, **c**, **d** symbol-charts of equidistant projections with negative and positive linear distortions, respectively, and **e** symbol-charts of equal-area projection

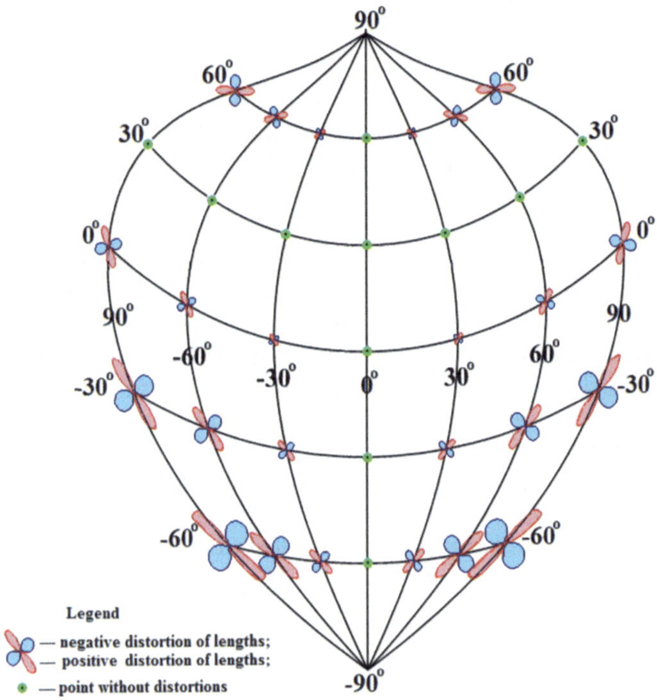

Fig. 1.19 Symbol-charts of distortions in the Bonn projection

Figure 1.19 displays the Bonn projection, the distortions of which are shown by using symbol-charts.

The symbol-charts can be used for more than just showing distortions, as will be discussed later in Sect. 1.7. They are capable of displaying any parameter that depends not only on the coordinates of the projection point but also on the direction.

References

Bagratuni GV (1962) Course in spheroidal geodesy, Technical Report FTD-MT-64-390, US Air Force, translation of Kurs sferoidicheskoi geodezii, 1967. (Geodezizdat, Moscow, 1962), by Foreign Technology Division, Wright-Patterson AFB, p 305. https://apps.dtic.mil/sti/pdfs/AD0650520.pdf

Bronstejn IN, Semendyaev KA, Nickel II, Marsolek L, Brunner J et al (2020) *Taschenbuch der Mathematik*. 11, aktualisierte Auflage. Haan-Gruiten, p 1280. ISBN 978-3-8085-5792-1

Bugayevskiy LM, Snyder JP (1995) Map projections. Reference Manual. Taylor &Francis Ltd., London, p 328. https://doi.org/10.1201/b16431

Kavraysky VV (1933) Issledovaniya po matematicheskoy kartografii (Studies in mathematical cartography). Trudy TSNIIGAiK, v 6, Leningrad, p 150. https://gis-lab.info/docs/tsniigaik-trudy/tsniigaik-trudy-006-1933.pdf

Kavraysky VV (1934) Matematicheskaya kartografiya (Mathematical cartography). Editorial Department of the ALLUNION Cartographic Trust. Moskow-Leningrad, p 276. https://elib.rgo.ru/handle/123456789/218456

Martínez-Llario JC, Baselga S, Coll E (2021) Accurate algorithms for spatial operations on the spheroid in a spatial database management system. Appl Sci 11(5129):21. https://doi.org/10.3390/app11115129

NGA.STND.0036_1.0.0_WGS84 (2014) National geospatial-intelligence agency (NGA) standardization document department of defense world geodetic system 1984. 2014–07–08 Version 1.0.0. Office of Geomatics. https://nsgreg.nga.mil/doc/view?i=4085

Papadopoulos A (2017) Nicolas-Auguste Tissot: a link between cartography and quasiconformal theory. HAL Id: hal-01406668. Arch Hist Exact Sci 71(4). https://doi.org/10.1007/s00407-016-0186-z. https://hal.archives-ouvertes.fr/hal-01406668

Rapp R (1991) Geometric geodesy. Part I. The Ohio State University Department of Geodetic Science and Surveying, Columbus, Ohio, p 189. https://geodesyattamucc.pbworks.com/f/Rapp_Geom_Geod_Vol_I.pdf

Rozhdestvensky BL, Yanenko NN (1978) Sistemy kvazilineynykh uravneniy i ikh prilozheniya k gazovoy dinamike (Systems of quasilinear equations and their applications to gas dynamics). Nauka, Moskow, p 688. https://libarch.nmu.org.ua/bitstream/handle/GenofondUA/69675/ab79de287eab1077172af6be65e8f205.djvu?sequence=1&isAllowed=y

Seif MR, Ghalenoei E (2018) Numerical determination of the geodesic curves: the solution of a two-point boundary value problem. Earth Observ Geomat Eng 2(1):26–35. https://doi.org/10.22059/eoge.2018.257564.1023. https://eoge.ut.ac.ir/article_66947_b329ce8007bd614cdd6f213f2e08792b.pdf

Serapinas BB (2005) Matematicheskaya kartografiya: Uchebnik dlya vuzov (Mathematical cartography: textbook for universities). Izdatel'skiy tsentr: Akademiya, Moskow, p 336. ISBN 5-7695-2131-7. https://www.geokniga.org/books/35973

Sjöberg LE (2012) Solutions to the ellipsoidal Clairaut constant and the inverse geodetic problem by numerical integration. J Geodetic Sci 2(3):162–171. https://doi.org/10.2478/v10156-011-0037-4

Snyder JP (1987) Map projections—a working manual, U.S. Geological Survey professional paper 1395, USPOGO, Washington D.C, p 397. https://doi.org/10.3133/pp1395

Tissot NA (1881) Mémoire sur la représentation des surfaces et les projections des cartes géographiques, Gauthier–Villars, Paris 1881, p 440. https://books.google.com.ua/books/about/Mémoire_sur_la_représentation_des_surf.html?id=eYWxzwEACAAJ&redir_esc=y

Chapter 2
The Problem of Finding the Best Projections

2.1 Euler–Urmaev System

Let us decipher the concept of "best" and "ideal" projection. To do this, we write expressions for particular extremal linear scale factors of an arbitrary homeomorphic projection of a regular surface onto a plane. Using formulae (1.87–1.90), we write:

$$\left.\begin{aligned} a^2 &= \tfrac{1}{2r^2}\left(e+g+\sqrt{(e+g)^2-4h^2}\right) \\ b^2 &= \tfrac{1}{2r^2}\left(e+g-\sqrt{(e+g)^2-4h^2}\right) \end{aligned}\right\}. \tag{2.1}$$

Obviously, a real ideal projection of a surface area onto a plane would be one for which the extremal scale factors at each point of the mapping area are equal to one, i.e.,

$$a = b = 1. \tag{2.2}$$

However, these equalities are not always possible since the unknown mapping functions N, E, on whose first derivatives the quantities e, g, h depend (see formulae (2.1)), must have continuous second mixed partial derivatives, i.e., the following equations must be satisfied:

$$N_{q\lambda} = N_{\lambda q}, E_{q\lambda} = E_{\lambda q}. \tag{2.3}$$

This means that there is a system of two equations connecting the functions a, b. Let us find this system.

From equalities (1.107–1.108) for projections with Jacobian $h \geq 0$, we obtain:

$$2ab = 2\frac{h}{r^2}. \tag{2.4}$$

E. Novikova, *Best Map Projections*, https://doi.org/10.1007/978-3-031-78334-0_2

Summing (2.1) and (2.4) and then subtracting (2.4) from (2.1), we obtain:

$$\begin{aligned} (a+b)r &= \sqrt{(N_q+E_\lambda)^2+(N_\lambda-E_q)^2} \\ (a-b)r &= \sqrt{(N_q-E_\lambda)^2+(N_\lambda+E_q)^2} \end{aligned}. \tag{2.5}$$

Let us introduce the following notations:

$$\tan\beta = \frac{N_\lambda - E_q}{N_q + E_\lambda}, \quad \tan\chi = \frac{N_\lambda + E_q}{N_q - E_\lambda}. \tag{2.6}$$

Then:

$$N_\lambda - E_q = r(a+b)\sin\beta, N_q + E_\lambda = r(a+b)\cos\beta. \tag{2.7}$$

$$N_\lambda + E_q = r(a-b)\sin\chi, N_q - E_\lambda = r(a-b)\cos\chi \tag{2.8}$$

From systems (2.7) and (2.8), we find $N_q, N_\lambda, E_q, E_\lambda$:

$$\begin{aligned} N_q &= \tfrac{r}{2}((a+b)\cos\beta + (a-b)\cos\chi) \\ N_\lambda &= \tfrac{r}{2}((a+b)\sin\beta + (a-b)\sin\chi) \end{aligned}, \tag{2.9}$$

$$\begin{aligned} E_q &= -\tfrac{r}{2}((a+b)\sin\beta - (a-b)\sin\chi) \\ E_\lambda &= \tfrac{r}{2}((a+b)\cos\beta - (a-b)\cos\chi) \end{aligned}. \tag{2.10}$$

After substituting the derivatives $N_q, N_\lambda, E_q, E_\lambda$ from (2.9–2.10) into (2.3), the continuity conditions for the second partial derivatives of the mapping functions N, E take the form:

$$\frac{\partial}{\partial\lambda}[r((a+b)\cos\beta + (a-b)\cos\chi)] = \frac{\partial}{\partial q}[r((a+b)\sin\beta + (a-b)\sin\chi)], \tag{2.11}$$

$$\frac{\partial\lambda}{\partial}\left[r((a+b)\sin\beta - (a-b)\sin\chi)\right] = -\frac{\partial}{\partial q}\left[r((a+b)\cos\beta - (a-b)\cos\chi)\right]. \tag{2.12}$$

The system of Eqs. (2.11–2.12) relates four projection characteristics: extremal scale factors a, b and angles β, χ. This system is called the **Euler–Urmaev system** (Bugayevskiy and Snyder 1995, p. 172).

Let us replace the angles β, χ in Eqs. (2.11) and (2.12) with the angles α_0, A_0, determined by the relations:

$$\alpha_0 = \frac{1}{2}(\beta + \chi), A_0 = \frac{1}{2}(\chi - \beta). \tag{2.13}$$

Let us show that α_0 is the azimuth of the main direction on the original surface and that A_0 is the grid azimuth of the main direction on the plane in the projection. Indeed:

$$\tan(\beta+\chi)=\frac{\tan\beta+\tan\chi}{1-\tan\beta\tan\chi},\ \tan(\chi-\beta)=\frac{\tan\chi-\tan\beta}{1+\tan\chi\tan\beta}, \tag{2.14}$$

or, considering (1.80, 1.99, 1.101, 2.6),

$$\tan(\beta+\chi)=\frac{2\left(N_qN_\lambda+E_qE_\lambda\right)}{N_q^2+E_q^2-N_\lambda^2-E_\lambda^2}=\frac{2f}{e-g}=\tan 2\alpha_0. \tag{2.15}$$

$$\tan(\chi-\beta)=\frac{2\left(N_qE_q+N_\lambda E_\lambda\right)}{N_q^2+N_\lambda^2-E_q^2-E_\lambda^2}=\frac{2f'}{e'-g'}=\tan 2A_0. \tag{2.16}$$

Hence:

$$\beta=\alpha_0-A_0,\ \chi=\alpha_0+A_0. \tag{2.17}$$

Thus, the angle β is the difference between the azimuth of the main direction on the surface of the ellipsoid and the grid azimuth of the main direction on the plane. Similarly, the angle χ is the sum of the azimuth of the main direction on the surface and the grid azimuth of the main direction on the plane.

Taking into account notation (2.13), formulae (2.11–2.12) can be rewritten as follows:

$$\frac{\partial}{\partial\lambda}[r(a\cos\alpha_0\cos A_0+b\sin\alpha_0\sin A_0)]=\frac{\partial}{\partial q}[r(a\sin\alpha_0\cos A_0-b\cos\alpha_0\sin A_0)], \tag{2.18}$$

$$\frac{\partial}{\partial\lambda}[r(a\cos\alpha_0\sin A_0-b\sin\alpha_0\cos A_0)]=\frac{\partial}{\partial q}[r(a\sin\alpha_0\sin A_0+b\cos\alpha_0\cos A_0)]. \tag{2.19}$$

Like system (2.11–2.12), the system of Eqs. (2.18–2.19) is an Euler–Urmaev system. Each variant of this system makes it possible to clearly show why it is impossible to simultaneously set the requirement that both extremal scale factors are equal to unity. Indeed, the use of conditions (2.2) in system (2.18–2.19) after identical transformations gives:

$$\tau_\lambda=\alpha_{0q}-A_{0q}=\beta_q, \tag{2.20}$$

$$\tau_q=-(\alpha_{0\lambda}-A_{0\lambda})=-\beta_\lambda. \tag{2.21}$$

Differentiating Eq. (2.20) with respect to λ and Eq. (2.21) with respect to q and summing, we obtain:

$$\Delta\tau = \tau_{qq} + \tau_{\lambda\lambda} = 0. \tag{2.22}$$

Therefore, for a projection to exist without distortion, the logarithm of the coefficient of the metric shape of the surface must be a harmonic function. Surfaces whose metric form coefficient satisfies condition (2.22) are discussed in detail by Grafarend and Krumm (2006, pp. 153–156).

In fact, the Laplace operator of the logarithm of the coefficient of the metric form of a surface is an arbitrary function depending on the properties of the surface and the chosen coordinate system (see Sect. 1.11). For an ellipsoid of revolution with the coordinate system q, λ:

$$\Delta\tau = -\frac{\cos^2\varphi\left(1 - e^2\sin^2\varphi\right)}{\left(1 - e^2\right)} < 0. \tag{2.23}$$

Everywhere, except for the points of the Pole, at which $\cos\varphi = 0$, the function $\Delta\tau$ is negative.

Therefore, the problem arises of finding a projection in which linear distortions would be minimal. This problem was first formulated by Kavraysky in (1934), who wrote that it is necessary to look for a projection with the sole condition that the largest linear distortion within the entire mapping area deviates from zero as little as possible. He called such projections "ideal".

2.2 Selecting a Linear Distortion Measure for Projection

Investigating the problem of finding ideal projections, Meshcheryakov (1961) showed that this problem cannot be defined unambiguously. To formulate this problem clearly, it is necessary to do the following:

1. Choose a measure of linear distortions at a given projection point in a specific direction;
2. Choose a general measure of linear distortions at a point;
3. Select a general measure of linear distortions within the mapping area.

Typically, one of the following values is taken as a measure of linear distortions at a projection point in a certain direction (Bugayevskiy and Snyder 1995, p. 26):

$$V = \mu - 1,\ V' = \ln\mu,\ V'' = 1 - \frac{1}{\mu},\ V''' = \frac{1}{2}\left(\mu^2 - 1\right). \tag{2.24}$$

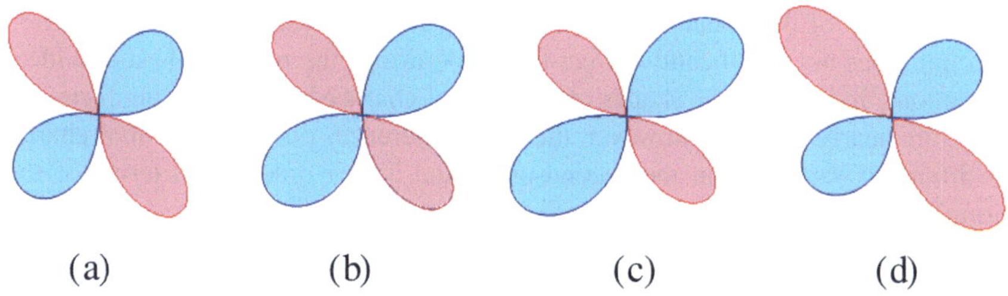

Fig. 2.1 Symbol-charts with coordinates, equal $\varphi = -30°, \lambda = -30°$, for the Bonn projection constructed using the following formulae: **a** $V = \mu - 1$; **b** $V' = \ln \mu$; **c** $V'' = 1 - \frac{1}{\mu}$; **d** $V''' = \frac{1}{2}(\mu^2 - 1)$

Figure 2.1 shows symbol-charts constructed using all four formulae of the system (2.24) for the same point with coordinates $\varphi = -30°, \lambda = -30°$ on the Bonn projection map grid (see Fig. 1.19).

The appearance and practical use of formulae for calculating linear distortions different from the original formula (1.134) can be justified by the following considerations:

1. During the eighteenth to twentieth centuries, cartographers and surveyors performed calculations using tables of logarithms. This is why the quantity V' has become widespread for assessing linear distortions in projections. Another argument for using the logarithmic formula to determine linear distortion is that in conformal projections, the function $\ln r + \ln \mu$ is the harmonic conjugate of the function $-\beta$. For more details, see Sect. 8.2.
2. The value V'' allows us to estimate the distortions of the lengths of not the direct but the inverse projection with the opposite sign. In our case, the inverse projection is a projection from a plane onto the surface of an ellipsoid. If the forward projection is the best, then the inverse projection is also the best.
3. The last formula of the system (2.24) does not use the linear scale factor itself but rather the square of this value. Let us return to Eq. (1.74), which determines the value μ^2. The linear scale factor can be found by taking the square root of the value μ^2. That is, the initial value is μ^2, not μ. . Therefore, it is easier to find linear distortions using the last formula of the system (2.24) and not using the formula (1.134).

Kavraysky (1934) showed that functions V, V', V'', V''' differ from each other in terms of quantities of the second order of smallness if these functions themselves are quantities of the first order of smallness. The conclusion is that we must use such a function (2.24), which leads to simpler mathematical calculations. For example, Tchebychev (1856) used the measure V' to find the best conformal projection. Markov in (1895) used not one measure but two. To limit the linear distortions from above, he chose the value V, and to limit the linear distortions from below, he chose the value V'' with the opposite sign. For more information about the Markov projection, see Sect. 8.3.

The choice of the linear distortion measure at the projection point in a certain direction does not significantly affect the properties of the resulting best and ideal projections. In Fig. 2.1, the shapes of the symbol-charts constructed using formulae (2.24) are nearly identical; however, there are differences due to the symbol-charts' sensitivity to variations in the second-order and higher-order small terms of the formulae.

There are an infinite number of formulae that allow us to calculate a value that differs from the value defined by formula (1.134) by a value of the second order of smallness if the value V is of the first order of smallness. However, the most natural and most commonly used measure of linear distortion is the function V.

2.3 Selecting a General Measure of Linear Distortions at a Point

To solve the problem of choosing a measure of linear distortions at a point, we need to answer the following question: How can distortions be compared at two different points of projection, if at each of these points, there are infinitely many different linear distortions?

To answer this question, it is necessary to replace all the variety of linear distortions at a point with one number, called the measure of linear distortions at a point. Different authors have solved the problem of finding the measure of linear distortions at a point in different ways. The German scientist V. Jordan (Bugayevskiy and Snyder 1995, p. 26) considered all possible linear distortions at the projection point and proposed the following integral value to determine the measure:

$$\varepsilon_J^2 = \frac{1}{2\pi}\int_0^{2\pi}(\mu - 1)^2 \mathrm{d}\alpha = \frac{1}{2\pi}\int_0^{2\pi} V^2 \mathrm{d}\alpha. \tag{2.25}$$

Integration in formula (2.25) is carried out in all directions, which makes it possible to consider all linear distortions at the projection point.

Another approach is to consider only the maximum and minimum linear distortions in the projection. In this case, at each projection point, two numbers are specified: the maximum and minimum linear distortions. However, even in this case, there is no unambiguous answer to the question of at which point the linear distortions will be lower and at which point more. For example, at one projection point, the following extremal linear distortions occur:

$$V_a = 3\%\ ,\ V_b = 1\%\ .$$

At another projection point, the extremal linear distortions are equal to

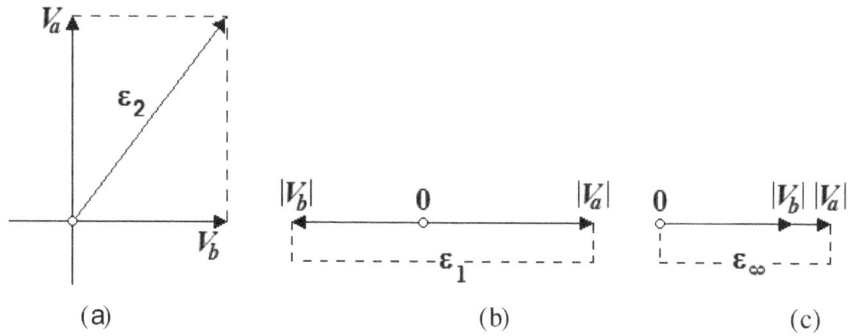

Fig. 2.2 Options for specifying the measure of length distortions at a point. **a** According to the Airy measure **b** According to the arithmetic mean measure. **c** According to the Chebyshev measure

$$V_a = V_b = 2\% \,.$$

At which point is the linear distortions less?

To answer this question, it is necessary to determine the measure of linear distortion at a point. For example, we can use the Cartesian coordinate system and plot the maximum linear distortion along one axis and the minimum distortion along the other axis, as shown in Fig. 2.2a. In this case, the measure can be defined as follows:

$$\varepsilon_A^2 = \left(V_q^2 + V_b^2\right). \tag{2.26}$$

This is exactly the formula for calculating the measure of linear distortions at a point that was proposed by Airy (1861).

The abovementioned method of taking into account linear distortions at a point on the basis of maximum and minimum distortions is not the only possible method. For example, we can plot the linear distortion values in absolute value on one numerical axis: V_a—to the right of zero; V_b—to the left, as shown in Fig. 2.2b.

The following formula can be used to calculate the linear distortion measure in this case:

$$\varepsilon_1 = |V_a| + |V_b|. \tag{2.27}$$

If $|V_a|$, $|V_b|$ are plotted on the number line on one side of zero, as shown in Fig. 2.2c, then the measure of linear distortions at the projection point can be determined by using the following formula:

$$\varepsilon_{\text{Ch}} = \varepsilon_\infty = \max(|V_a|, |V_b|). \tag{2.28}$$

This quantity is called the Chebyshev measure, or the measure of uniform approximation (Bugayevskiy and Snyder 1995, p. 193).

The problem of choosing a measure at a point under consideration corresponds to the problem of choosing a metric in a space consisting of two numbers (Korn and Korn 1968). Using the general approach of entering a metric in the space of two numbers, we write:

$$\varepsilon_j = \left(|V_a|^j + |V_b|^j\right)^{\frac{1}{j}}, \tag{2.29}$$

where $j \geq 1$.

For $j = 1$, Eq. (2.27) is obtained; for $j = 2$, Eq. (2.26) is obtained; and for $j \to \infty$, Eq. (2.28) is obtained.

A unique measure of linear distortions at a point was proposed by A. Markov. This measure is based on Chebyshev's concept but has features associated with the Markov projection (see Sect. 8.3). The Markov measure is calculated either by the maximum scale factor:

$$V_{M\_a} = a - 1, \tag{2.30}$$

or by the minimum scale factor:

$$V_{M\_b} = \frac{1}{b} - 1. \tag{2.31}$$

From two values calculated using formulae (2.30–2.31), the maximum value is selected:

$$\varepsilon_M = \max\left(V_{M\_a}, V_{M\_b}\right). \tag{2.32}$$

A. Markov considers the value ε_M to be the Chebyshev measure characterizing the linear distortions in the projection and finds the best conical projection using this measure. If the projection linear distortions are quantities of the first order of smallness, then the value ε_M, up to quantities of the second order of smallness, is equal to the Chebyshev measure.

2.4 Selecting a Criterion of Linear Distortions for the Entire Mapping Area

The next step is to introduce a criterion that allows one to evaluate linear distortions not at one projection point but throughout the entire mapping area. There is no single approach to this, and at least two methods can be used to assess distortions.

The first method considers all the measures of linear distortions at each projection point. In this case, the criterion for assessing linear distortions is as follows:

$$\Xi_j = \left(\frac{1}{\sigma}\int \varepsilon_j^t d\sigma\right)^{\frac{1}{t}}. \tag{2.33}$$

In formula (2.33), integration is carried out over the mapping area σ. As in the previous case, the value $t \geq 1$. The criterion represented by formula (2.33) is called a variational type of criterion.

As a rule, the degree of the linear distortion measure at a point and the degree of the criterion are chosen to be the same, i.e., if $j = 2$, then $t = 2$. In this case, Eq. (2.33) is rewritten as follows:

$$\Xi_j = \left(\frac{1}{\sigma}\int\limits_{\sigma} (\varepsilon_j)^j d\sigma\right)^{\frac{1}{j}}. \tag{2.34}$$

For $j = 2$, formula (2.34) gives a criterion known as the Airy criterion. If the Jordan measure is used in formula (2.34) instead of ε_j (formula (2.25)), then the Jordan criterion will be obtained. Other variants of the criterion can be found in (Bugayevskiy and Snyder 1995, p. 26).

The second method for assessing projection distortions considers only the maximum value of distortions for the entire mapping area, defined as:

$$\Xi_{\text{Ch}} = \max_{\sigma}(|V_a|, |V_b|). \tag{2.35}$$

This criterion is called Chebyshev's criterion (Bugayevskiy and Snyder 1995, p. 193).

Notably, Chebyshev's criterion is related to the variational type of criterion by the following relationship:

$$\Xi_{\text{Ch}} = \lim_{j\to\infty} \Xi_j. \tag{2.36}$$

As mentioned above, a unique version of Chebyshev's criterion was proposed by Markov (1895). According to this criterion, a projection is selected for which the condition is satisfied:

$$\Xi_{\text{Ch}-M} = \max_{\sigma}(\varepsilon_M) = \min. \tag{2.37}$$

The best projections according to any of the above criteria, which are selected from the entire set of twice differentiable projections, are called **ideal projections** according to this criterion. The projections that are the best according to some criterion in a limited set, for example, among all conformal ones, are called the **best projections** in this set. An example is the best Chebyshev projection, which is the best among only conformal projections.

References

Airy GB (1861) Explanation of a projection by balance of errors for maps applying to a very large extent of the Earth's surface and comparison of this projection with other projections. London, Edinburgh, and Dublin Philosoph Mag 4, 22(149):409–421. https://doi.org/10.1080/14786446108643179

Bugayevskiy LM, Snyder JP (1995) Map projections. Reference Manual. Taylor&Francis Ltd., London, p 328. https://doi.org/10.1201/b16431

Grafarend EW, Krumm FW (2006) Map projections. Cartographic information systems. Springer-Verlag Berlin Heidelberg, p 713. https://books.google.com.ua/books/about/Map_Projections.html?id=Rg3Lc8omIXMC&redir_esc=y

Kavraysky VV (1934) Matematicheskaya kartografiya (Mathematical cartography). Editorial Department of the ALLUNION Cartographic Trust. Moskow-Leningrad, p 276. https://elib.rgo.ru/handle/123456789/218456

Korn GA, Korn TM (1968) Mathematical handbook: for scientists and engineers. McGraw-Hill Book Company. New York, San Francisco, Toronto, London, Sydney, p 832. https://archive.org/details/mathematicalhand0000gran/page/n7/mode/2up

Markov AA (1895) O naivygodneyshikh izobrazheniyakh nekotoroy chasti dannoy poverkhnosti vrashcheniya na ploskosti (On the most advantageous images of a certain part of a given surface of rotation on a plane), Izvestiya Imperatorskoy Akademii Nauk 2(3):177–187. https://www.mathnet.ru/rus/im4207

Meshcheryakov GA (1961) New extremal problems. Dokl. Akad. Nauk SSSR 136(5):026–1029. http://www.mathnet.ru/eng/agreement

Tchebychev P (1856) Sur la construction des cartes géographiques. Notes 13. (Lu le 18 janvier 1856). Bulletin de la Classe Phisico-Mathématique de L'Académie Impériale des Sciences de Saint-Pétersbourg T. XIV. 17(529):257–261. https://www.biodiversitylibrary.org/item/173056#page/629/mode/1up

Chapter 3
Ideal Projection According to the Airy Criterion

3.1 Ideal Projection Theorem

The search for an ideal projection will be conducted using the Airy criterion, i.e., based on the condition (Airy 1861; Bugayevskiy and Snyder 1995, p. 26; Grafarend and Krumm 2006, p. 86):

$$\Xi_A^2 = \frac{1}{2\sigma}\iint\left[(a-1)^2 + (b-1)^2\right]\mathrm{d}\sigma = \min. \tag{3.1}$$

Taking into account equalities (1.102, 2.5), formula (3.1) can be represented in the following form:

$$\Xi_A^2 = \frac{1}{2\sigma}\iint\left(N_q^2 + E_q^2 + N_\lambda^2 + E_\lambda^2 - 2r\sqrt{(N_q+E_\lambda)^2 + (N_\lambda - E_q)^2} + 2r^2\right)\mathrm{d}q\mathrm{d}\lambda. \tag{3.2}$$

As is known (Korn and Korn, p. 355), if the functional Ξ^2 reaches a minimum on the functions N, E, then they satisfy the system of Euler–Ostrogradsky differential equations:

$$\frac{\partial Q}{\partial N} - \frac{\partial}{\partial q}\left(\frac{\partial Q}{\partial N_q}\right) - \frac{\partial}{\partial \lambda}\left(\frac{\partial Q}{\partial N_\lambda}\right) = 0,\ \frac{\partial Q}{\partial E} - \frac{\partial}{\partial q}\left(\frac{\partial Q}{\partial E_q}\right) - \frac{\partial}{\partial \lambda}\left(\frac{\partial Q}{\partial E_\lambda}\right) = 0, \tag{3.3}$$

and natural boundary conditions:

$$\frac{\partial Q}{\partial N_q}\cdot\frac{\partial \lambda}{\partial s} - \frac{\partial Q}{\partial N_\lambda}\cdot\frac{\partial q}{\partial s} = 0,\ \frac{\partial Q}{\partial E_q}\cdot\frac{\partial \lambda}{\partial s} - \frac{\partial Q}{\partial E_\lambda}\cdot\frac{\partial q}{\partial s} = 0, \tag{3.4}$$

E. Novikova, *Best Map Projections*, https://doi.org/10.1007/978-3-031-78334-0_3

where Q is the integrand in formula (3.2) and ∂s is the differential of the arc of the boundary of the region.

Let us find the derivatives included in systems (3.3–3.4).

$$\begin{aligned}
&\frac{\partial Q}{\partial N}=0,\ \frac{\partial Q}{\partial E}=0\\
&\frac{\partial Q}{\partial N_q}=2\left(N_q-\frac{r(N_q+E_\lambda)}{\sqrt{(N_q+E_\lambda)^2+(N_\lambda-E_q)^2}}\right)\\
&\frac{\partial Q}{\partial N_\lambda}=2\left(N_\lambda-\frac{r(N_\lambda-E_q)}{\sqrt{(N_q+E_\lambda)^2+(N_\lambda-E_q)^2}}\right)\\
&\frac{\partial Q}{\partial E_q}=2\left(E_q+\frac{r(N_\lambda-E_q)}{\sqrt{(N_q+E_\lambda)^2+(N_\lambda-E_q)^2}}\right)\\
&\frac{\partial Q}{\partial E_\lambda}=2\left(E_\lambda-\frac{r(N_q+E_\lambda)}{\sqrt{(N_q+E_\lambda)^2+(N_\lambda-E_q)^2}}\right)
\end{aligned}\tag{3.5}$$

Substituting (3.5) into (3.3), taking into account notation (2.7), after identical transformations leads to:

$$\begin{aligned}
&\tfrac{\partial}{\partial q}\left(N_q-r\cos\beta\right)+\tfrac{\partial}{\partial\lambda}(N_\lambda-r\sin\beta)+E_{\lambda q}-E_{q\lambda}=0\\
&\tfrac{\partial}{\partial q}(E_q+r\sin\beta)+\tfrac{\partial}{\partial\lambda}(E_\lambda-r\cos\beta)+N_{\lambda q}-N_{q\lambda}=0
\end{aligned},\tag{3.6}$$

or,

$$\begin{aligned}
&\tfrac{\partial}{\partial q}\left[r(a+b-1)\cos\beta\right]=-\tfrac{\partial}{\partial\lambda}\left[r(a+b-1)\sin\beta\right]\\
&\tfrac{\partial}{\partial q}\left[r(a+b-1)\sin\beta\right]=\tfrac{\partial}{\partial\lambda}\left[r(a+b-1)\cos\beta\right]
\end{aligned}.\tag{3.7}$$

Analyzing system (3.7), we conclude that $r(a+b-1)\cos\beta$ and $-r(a+b-1)\sin\beta$ are conjugate harmonic functions. System (3.7) can be represented in the following form:

$$\Phi_q=-\beta_\lambda,\ \Phi_\lambda=\beta_q,\tag{3.8}$$

where

$$\Phi=\ln r(a+b-1).\tag{3.9}$$

Consequently, $\ln r(a+b-1)$, and $(-\beta)$ are conjugate harmonic functions.

Taking into account (2.9, 2.10, 3.5), the natural boundary conditions are written as follows:

$$\begin{aligned}
[(a+b-2)\cos\beta + (a-b)\cos\chi]\tfrac{\partial}{\partial s} &= [(a+b-2)\sin\beta + (a-b)\sin\chi]\tfrac{\partial q}{\partial s} \\
[-(a+b-2)\sin\beta + (a-b)\sin\chi]\tfrac{\partial}{\partial s} &= [(a+b-2)\cos\beta - (a-b)\cos\chi]\tfrac{\partial q}{\partial s}
\end{aligned}. \quad (3.10)$$

Dividing the first condition of the system (3.10) by the second, we have:

$$(a+b-2)^2 = (a-b)^2. \quad (3.11)$$

Taking the roots from the right and left sides of Eq. (3.11), we obtain:

$$|a+b-2| = a-b. \quad (3.12)$$

If at the border:

$$a+b-2 \geq 0, \quad (3.13)$$

then Eq. (3.12) can be rewritten as follows:

$$b = 1. \quad (3.14)$$

If at the border

$$a+b-2 \leq 0, \quad (3.15)$$

then:

$$a = 1. \quad (3.16)$$

Let inequality (3.13) be satisfied at the boundary. The first equation of system (3.10) can be written as follows:

$$\frac{\sin\beta + \sin\chi}{\cos\beta + \cos\chi} = \frac{\partial}{\partial s}\Big/\frac{\partial q}{\partial s}, \quad (3.17)$$

or,

$$\tan\frac{\beta+\chi}{2} = \frac{\partial}{\partial s}\Big/\frac{\partial q}{\partial s}. \quad (3.18)$$

In the right part of Eq. (3.18), the value of the tangent of the azimuth of the tangent line to the boundary of the region is given, whereas in the left part, the tangent of the azimuth of the main direction of the projection is provided (see Eq. 2.13). Therefore, taking into account (3.18), we obtain:

$$\alpha_0 = \alpha_s + \pi j, \ (j = 0, 1), \quad (3.19)$$

where α_s is the azimuth of the tangent to the boundary of the region.

If inequality (3.15) is satisfied at the boundary, then the solution is similar to that above.

As indicated in Sect. 1.6.3, at each projection point, there are two mutually perpendicular main directions. By analyzing formula (3.19), we conclude that for an ideal projection, one of them coincides with the tangent, and the other coincides with the normal to the boundary of the region.

Therefore, the above formulae prove the following theorem (Novikova 1989, 1991):

An ideal projection according to the Airy criterion is a projection in which $\ln r(a+b-1)$ and $(-\beta)$ are conjugate harmonic functions. At the boundary of the mapping area, one of the extremal scale factors is equal to unity, and Tissot's indicatrixes are oriented along tangents or normals to the boundary of the region.

3.2 Relationships Between Mapping Functions and Green's Function

From system (3.6), we find the Laplace operators of the functions N and E.

$$\begin{aligned}\Delta N &= \tfrac{\partial}{\partial q}(r\cos\beta) + \tfrac{\partial}{\partial \lambda}(r\sin\beta)\\ \Delta E &= \tfrac{\partial}{\partial \lambda}(r\cos\beta) - \tfrac{\partial}{\partial q}(r\sin\beta)\end{aligned}. \tag{3.20}$$

Let $\overline{N}, \overline{E}$ be harmonic functions coinciding on the boundary with N, E. Then (Korn and Korn 1968, p. 480; Timan and Trofimov 1968):

$$\begin{aligned}N(A) &= -\iint G(A)\Big(\tfrac{\partial}{\partial q}(r\cos\beta) + \tfrac{\partial}{\partial \lambda}(r\sin\beta)\Big)\mathrm{d}q\mathrm{d}\lambda + \overline{N}(A)\\ E(A) &= -\iint G(A)\Big(\tfrac{\partial}{\partial \lambda}(r\cos\beta) - \tfrac{\partial}{\partial q}(r\sin\beta)\Big)\mathrm{d}q\mathrm{d}\lambda + \overline{E}(A)\end{aligned}, \tag{3.21}$$

where $G(A)$ is the Green's function generated for a point A located inside the mapping area. By applying the integration by parts to the integrals and taking into account that the Green's function is equal to zero at the boundary of the region, we obtain:

$$\begin{aligned}N(A) &= -\iint r\Big(\tfrac{\partial G(A)}{\partial q}\cos\beta + \tfrac{\partial G(A)}{\partial \lambda}\sin\beta\Big)\mathrm{d}q\mathrm{d}\lambda + \overline{N}\\ E(A) &= -\iint r\Big(\tfrac{\partial G(A)}{\partial \lambda}\cos\beta - \tfrac{\partial G(A)}{\partial q}\sin\beta\Big)\mathrm{d}q\mathrm{d}\lambda + \overline{E}\end{aligned}. \tag{3.22}$$

Let us introduce the notation:

$$\tan\varsigma = \frac{\partial G(A)}{\partial \lambda} \Big/ \frac{\partial G(A)}{\partial q}. \tag{3.23}$$

The system (3.22) is then rewritten as follows:

$$\begin{matrix} N = \iint r|\nabla G(A)| \cos \psi \mathrm{d}q\mathrm{d}\lambda + \overline{N} \\ E = \iint r|\nabla G(A)| \sin \psi \mathrm{d}q\mathrm{d}\lambda + \overline{E} \end{matrix}, \tag{3.24}$$

where

$$|\nabla G| = |\mathrm{grad}G| = \sqrt{\left(\frac{\partial G}{\partial q}\right)^2 + \left(\frac{\partial G}{\partial \lambda}\right)^2}, \tag{3.25}$$

$$\psi = \varsigma - \beta. \tag{3.26}$$

In Eq. (3.24), ∇G is the gradient of the Green's function; ψ is the difference between the azimuth of the gradient vector of the Green's function and the angle β.

3.3 Properties of an Ideal Airy Projection

3.3.1 Distortions of Angles and Areas of Ideal Projection

Let us transform the Airy measure as follows:

$$\varepsilon_2^2 = \left((a-1)^2 + (b-1)^2\right) = \frac{1}{2}\left((a+b-2)^2 + (a-b)^2\right). \tag{3.27}$$

If the linear distortions in the projection are quantities of the first order of smallness, then, taking into account equalities (1.135, 1.147, 1.151), we can write:

$$\varepsilon_2^2 = \frac{1}{2}\left(\omega^2 + V_p^2\right). \tag{3.28}$$

From this, we can conclude that the resulting ideal projection minimizes not only linear distortions but also angular and area distortions.

3.3.2 Linear Scale Factors at the Boundary of the Region

Since the function $\ln r(a+b-1)$ is harmonic, it satisfies Laplace's equation:

$$\Delta \ln r(a+b-1) = \Delta \ln(a+b-1) + \Delta \ln r = 0. \tag{3.29}$$

As the function $\Delta \ln r$ retains its sign in the mapping area, the function $\ln(a + b - 1)$ is either subharmonic or superharmonic (Timan and Trofimov 1968, p. 180).

In an ideal projection, all distortions tend to a minimum. Consequently, the subharmonic (superharmonic) function $\ln(a + b - 1)$ should also tend to a minimum. If the linear distortions in the projection are quantities of the first order of smallness, then the subharmonic (superharmonic) function $\ln(a + b - 1)$ satisfies Grave's lemma (Grave 1911), i.e., on the border:

$$\ln(a + b - 1) \approx H = \text{const.} \tag{3.30}$$

If $\Delta \ln r \leq 0$, then H is the positive maximum of the function $\ln(a + b - 1)$. The function $\ln(a + b - 1)$ reaches a minimum equal to $(-H)$ at some point inside the region.

Then, at the boundary:

$$a + b - 1 \approx \exp H > 1. \tag{3.31}$$

Therefore,

$$a \approx \exp H > 1, \quad b = 1. \tag{3.32}$$

Similarly, if $\Delta \ln r \geq 0$, then at the boundary:

$$a = 1, b \approx \exp H \leq 1. \tag{3.33}$$

The requirement that the function $\ln(a + b - 1)$ should deviate least from zero is not necessary when finding an Airy ideal projection. Nevertheless, for an approximate calculation of the ideal projection, it is convenient to either replace one of the boundary conditions (3.12), (3.19) with condition (3.33) or use this condition along with the boundary conditions (3.12), (3.19). Otherwise, to find the ideal projection, it is necessary to use the method of successive approximations.

3.3.3 *Area Distortion*

If the linear distortions of the projection are quantities of the first order of smallness, then taking into account (1.147) we obtain:

$$V_p \approx \ln(a + b - 1). \tag{3.34}$$

Consequently, in an ideal Airy projection, the area distortion is a function that is almost identical to the subharmonic (superharmonic) function $\ln(a + b - 1)$, which deviates the least from zero.

3.3.4 Geometric Meaning of Angle β

According to Eq. (2.17), the angle β is the difference between the azimuth of the main direction on the surface of the ellipsoid and the grid azimuth of the main direction of the plane.

Consequently, in an ideal Airy projection, the difference between the azimuth of the main direction of the displayed surface and the grid azimuth of the main direction on the plane is a harmonic function conjugated to the function $\ln r(a + b - 1)$.

3.3.5 Projection Center Point

Let us show the following property of an ideal projection: if functions a and b reach an extremum at one point inside the mapping area, then at this point, there are no distortions of angles, i.e.,

$$a = b. \tag{3.35}$$

Indeed, at the extremum point inside the mapping area, the following relations must be satisfied:

$$a_q = a_\lambda = b_q = b_\lambda = 0. \tag{3.36}$$

Taking into account (3.36), Eq. (3.7) becomes the following;

$$\tau_q = -\beta_\lambda, \quad \tau_\lambda = \beta_q. \tag{3.37}$$

Substituting (3.36, 3.37) into the Euler–Urmaev system (2.11–2.12) yields:

$$\begin{aligned}(a-b)\left(\tau_\lambda - \chi_q\right)\cos 2\alpha_0 - (a-b)\left(\tau_q + \chi_\lambda\right)\sin 2\alpha_0 = 0\\ (a-b)\left(\tau_\lambda - \chi_q\right)\sin 2\alpha_0 + (a-b)\left(\tau_q + \chi_\lambda\right)\cos 2\alpha_0 = 0\end{aligned}. \tag{3.38}$$

Solving this system for $(a - b)$, we obtain Eq. (3.35).

3.3.6 Range of Change of the Function $\ln(a + b - 1)$

As shown above, extremal scale factors at the central point reach an extremum. The function $\ln(a + b - 1)$ also has an extremum at this point. Since this function is subharmonic for the surface of an ellipsoid, it has a minimum at this point. Let us denote it as $\ln(a + b - 1)_{\min}$.

At the boundary of the region, the function $\ln(a + b - 1)$ reaches a maximum, equal to H (see Eq. (3.30)).

According to (Korn and Korn 1968, p. 480; Timan and Trofimov 1968, p. 70), a twice continuously differentiable function in an enclosed area at some point A inside the area is equal to:

$$F(A) = \iint G(A)\Delta F \mathrm{d}\sigma + \int F \frac{\mathrm{d}G(A)}{\mathrm{d}n} \mathrm{d}s. \tag{3.39}$$

The first integral of Eq. (3.39) is an integral over the mapping region, and the second integral is an integral over the boundary of the mapping region.

In formula (3.39), ΔF is the Laplace operator of the function F; $G(A)$ is the Green's function constructed for the Laplace operator. The Green's function is constructed individually for a given point in the region, meaning that each point has its own Green's function. Therefore, the point A in the parentheses of the Green's function in formula (3.39) indicates the point for which this function is constructed.

For the Laplace operator, the Green's function is determined by the following conditions (Timan and Trofimov 1968):

- The Green's function is harmonic, i.e., it satisfies Laplace's equation.
- Inside the definition area, the Green's function is negative.
- At the boundary of the region, the Green's function satisfies the following relations:

$$G(A) = 0, \int \frac{\mathrm{d}G(A)}{\mathrm{d}n} \mathrm{d}s = 1. \tag{3.40}$$

For the function $\ln(a + b - 1)$, according to Eq. (3.29), the Laplace operator is equal to:

$$\Delta \ln(a + b - 1) = -\Delta \ln r. \tag{3.41}$$

Taking into account (3.40, 3.41) and (3.39), formula (3.39) for the function $\ln(a + b - 1)$ is written as follows:

$$\ln(a + b - 1) = -\iint G(A)\Delta \ln r \mathrm{d}\sigma + \int \ln(a + b - 1)_{\max} \frac{\mathrm{d}G(A)}{\mathrm{d}n} \mathrm{d}s. \tag{3.42}$$

Let us select a point A_0 inside the mapping region at which the function $\ln(a + b - 1)$ reaches the minimum. Then, formula (3.43) can be written as follows:

$$\ln(a + b - 1)_{\max} - \ln(a + b - 1)_{\min} = \iint G(A_0)\Delta \ln r \mathrm{d}\sigma. \tag{3.43}$$

Thus, the range of variation of the function $\ln(a + b - 1)$ depends on the size of the mapping region and on the configuration of the boundary of this region.

Taking into account Eq. (3.30) and Grave's lemma (see Sect. 8.2), Eq. (3.43) can be written as follows:

$$H \approx \frac{1}{2} \iint G(A_0) \Delta \ln r d\sigma. \tag{3.44}$$

References

Airy GB (1861) Explanation of a projection by balance of errors for maps applying to a very large extent of the Earth's surface and comparison of this projection with other projections. London, Edinburgh, and Dublin Philosophical Magazine 4, 22(149):409–421. https://doi.org/10.1080/14786446108643179

Bugayevskiy LM, Snyder JP (1995) Map projections. Reference Manual. Taylor&Francis Ltd., London, p 328. https://doi.org/10.1201/b16431

Grafarend EW, Krumm FW (2006) Map Projections. Cartographic Information Systems. Springer-Verlag Berlin Heidelberg, p 713. https://books.google.com.ua/books/about/Map_Projections.html?id=Rg3Lc8omIXMC&redir_esc=y

Gravé D (1911) Demonstration d'un theoreme de Tchebychef generalize. Journal fur die reine und angewandte Mathematik. Band 140. Heft 4, 34:247–251. http://resolver.sub.uni-goettingen.de/purl?PPN243919689_0140

Korn GA, Korn TM (1968) Mathematical Handbook: for Scientists and Engineers. McGraw-Hill Book Company. New York, San Francisco, Toronto, London, Sydney. p 832. https://archive.org/details/mathematicalhand0000gran/page/n7/mode/2up

Novikova EN (1989) Ideal'naya po kriteriyu Eyri i nailuchshaya iz semeystva blizkikh k ravnovelikim proyektsii dlya obshirnykh oblastey proizvol'nykh ochertaniy (Ideal according to the Airy criterion and the best of the family of close-to- equal area projections for large areas of arbitrary shape). Krivoy Rog, p 53. Bibliogr. 14 nazv. Rus. Dep. v UkrNIINTI, 8.06.1989, N 1567 Uk-89Dep

Novikova EN (1991) Ideal'naya po kriteriyu Eyri proyektsiya dlya Sovetskogo Soyuza (Airy ideal projection for the Soviet Union). Izvestiya vysshih uchebnyh zavedeniy: Geodeziya i Aerofotosiemka. N 6. https://eposlink.com/ru/catalog/library/elibrary/book/izvestiya_vysshih_uchebnyh_zavedeniy_geodeziya_i_aerofotosemka-2290/publication/82141/

Timan AF, Trofimov VN (1968) Vvedeniye v teoriyu garmonicheskikh funktsiy (Introduction to the theory of harmonic functions). Nedra, Moskow, p 208

Chapter 4
The Best Projection from a Set of Close-to-Equal-Area Projections

4.1 Brief Historical Overview

All distortions of ideal projections are minimized, but the distortions of areas can be significant. Area distortions reach their extremum at the central point, where there are no angular distortions, and at the border, where angular distortions also reach their extremum. A.B. Peters (1984) noted that large area distortions on world maps cause misconceptions about the size of continents. The corresponding sizes of states, oceans, and continents can be realistically shown only on projections of equal areas and close to them. Therefore, the requirement of equal areas for small-scale maps remains preferable.

The problem of finding the best equal-area projections has a long history. Tsinger (1916) found the best equal-area conic projection by Airy criterion using the least squares method. Tsinger (1898) and V.V. Kavraysky (1933, 1934) hypothesized that the best equal-area projection is the one for which the linear scale factor is equal to some constant at the boundary of the region, i.e., they assumed that Chebyshev's theorem holds for equal-area projections. V. Hojovec (1984) used the least squares method to find the best equal-area projection by Airy criterion within a bounded set. J. Snyder (1988) showed by numerical methods that the best equal-area projection is among the projections that satisfy Chebyshev's theorem. Deakin (1990) chose an equal-area projection with minimal linear distortion among pseudocylindrical map projections. In the twenty-first century, the problem of finding the best projection among equal-area ones has also remained a priority in the scientific cartographic community (Canters and Crols 2011; Cheng and Lorre 2000).

An interesting method for finding the best equal-area projections was proposed by A.B. Peters (1984). In this method, infinitesimal quantities are replaced by finite quantities. A finite number of points are selected on the Earth's surface; if possible, on continents, all distances between these points are calculated using a high-speed computer, and then some constants of projection are determined, provided that these

E. Novikova, *Best Map Projections*, https://doi.org/10.1007/978-3-031-78334-0_4

distances are least distorted. As shown below, the problem of finding the best projection in a set of close-to-equal-area is solved quite simply and does not require the use of high-speed computers.

4.2 The Theorem on the Best Projection

Let us consider a set of projections for which the condition:

$$a + b - 2 = V_a + V_b = 0, \tag{4.1}$$

is satisfied.

The area distortions of these projections, according to formulae (1.135), are equal:

$$V_p = \text{ab} - 1 = (1 + V_a)(1 + V_b) - 1 = -V_a^2. \tag{4.2}$$

The area distortions of these projections are quantities of the second order of smallness if the linear distortions are quantities of the first order of smallness. Therefore, this set of projections can be called a set of close-to-equal-area projections.

The Airy criterion (Airy 1861) for these projections, taking into account formulae (3.1, 3.27, 4.1), is written as follows:

$$\Xi_A^2 = \frac{1}{2\sigma} \iint \left[(a-1)^2 + (b-1)^2\right] d\sigma = \frac{1}{4\sigma} \iint (a-b)^2 r^2 \text{dqd}\lambda. \tag{4.3}$$

The problem of finding the best projection in this set of projections is an optimal control problem with constraint (4.1) of the type of equality (Alekseev et al. 1979; Korn and Korn (1968), p. 355). Let us compose the Lagrange function:

$$Q' = Q + 4\text{Lr}^2(a + b - 2), \tag{4.4}$$

where Q is the integrand in (4.3) and where L is a certain function called the Lagrange multiplier.

Substituting Q from (4.3) into (4.4) yields:

$$Q' = r^2(a-b)^2 + 4\text{Lr}^2(a + b - 2), \tag{4.5}$$

or using (2.5),

$$Q' = (N_q - E_\lambda)^2 + (N_\lambda + E_q)^2 + 4\text{Lr}\left[\sqrt{(N_q + E_\lambda)^2 + (N_\lambda - E_q)^2} - 2r\right]. \tag{4.6}$$

As is known (Alekseev et al. 1979), if the functions N, E provide a minimum to the functional (4.3) under constraint (4.1), then the conditions for the Lagrange function to be stationary with respect to N and E look like this:

$$\begin{gathered}\frac{\partial Q'}{\partial N}-\frac{\partial}{\partial q}\left(\frac{\partial Q'}{\partial N_q}\right)-\frac{\partial}{\partial \lambda}\left(\frac{\partial Q'}{\partial N_\lambda}\right)=0 \\ \frac{\partial Q'}{\partial E}-\frac{\partial}{\partial q}\left(\frac{\partial Q'}{\partial E_q}\right)-\frac{\partial}{\partial \lambda}\left(\frac{\partial Q'}{\partial E_\lambda}\right)=0\end{gathered}. \tag{4.7}$$

Additionally, natural boundary conditions are satisfied at the boundary, namely:

$$\begin{gathered}\frac{\partial Q'}{\partial N_q}\cdot\frac{\partial \lambda}{\partial s}-\frac{\partial Q'}{\partial N_\lambda}\cdot\frac{\partial q}{\partial s}=0 \\ \frac{\partial Q'}{\partial E_q}\cdot\frac{\partial \lambda}{\partial s}-\frac{\partial Q'}{\partial E_\lambda}\cdot\frac{\partial q}{\partial s}=0\end{gathered}. \tag{4.8}$$

Let us find the derivatives included in equalities (4.7, 4.8).

$$\begin{gathered}\frac{\partial Q'}{\partial N}=0,\ \frac{\partial Q'}{\partial E}=0 \\ \frac{\partial Q'}{\partial N_q}=2\left[\left(N_q-E_\lambda\right)+\frac{2Lr\left(N_q+E_\lambda\right)}{\sqrt{(N_q+E_\lambda)^2+(N_\lambda-E_q)^2}}\right] \\ \frac{\partial Q'}{\partial N_\lambda}=2\left[\left(N_\lambda+E_q\right)+\frac{2Lr\left(N_\lambda-E_q\right)}{\sqrt{(N_q+E_\lambda)^2+(N_\lambda-E_q)^2}}\right] \\ \frac{\partial Q'}{\partial E_q}=2\left[\left(N_\lambda+E_q\right)-\frac{2Lr\left(N_\lambda-E_q\right)}{\sqrt{(N_q+E_\lambda)^2+(N_\lambda-E_q)^2}}\right] \\ \frac{\partial Q'}{\partial E_\lambda}=2\left[-\left(N_q-E_\lambda\right)+\frac{2Lr\left(N_q+E_\lambda\right)}{\sqrt{(N_q+E_\lambda)^2+(N_\lambda-E_q)^2}}\right]\end{gathered}. \tag{4.9}$$

Substituting (4.9) into (4.7), taking into account (2.7, 2.8, 4.1), leads to:

$$\begin{gathered}\frac{\partial}{\partial q}(N_q-E_\lambda+2Lr\cos\beta)+\frac{\partial}{\partial\lambda}(N_\lambda+E_q+2Lr\sin\beta)+2E_q-2E_q=0 \\ \frac{\partial}{\partial q}\left(N_\lambda+E_q-2Lr\sin\beta\right)+\frac{\partial}{\partial\lambda}(-N_q+E_\lambda+2Lr\cos\beta)+2N_q-2N_q=0\end{gathered}. \tag{4.10}$$

Under condition (4.1), formulae (2.7) can be rewritten as follows:

$$N_q+E_\lambda=2r\cos\beta,\quad N_\lambda-E_q=2r\sin\beta. \tag{4.11}$$

Considering (4.11), the system (4.10) transforms into:

$$\begin{gathered}\frac{\partial}{\partial q}(r(1+L)\cos\beta)=-\frac{\partial}{\partial\lambda}(r(1+L)\sin\beta) \\ \frac{\partial}{\partial q}(r(1+L)\sin\beta)=\frac{\partial}{\partial\lambda}(r(1+L)\cos\beta)\end{gathered}. \tag{4.12}$$

Analyzing system (4.12), we conclude that the functions $r(1+L)\cos\beta$ and $-r(1+L)\sin\beta$ are conjugate harmonic functions.

System (4.12) can be represented as:

$$\Phi'_q=-\beta_\lambda,\ \Phi'_\lambda=\beta_q, \tag{4.13}$$

where

$$\Phi\prime = \ln r(1 + L). \tag{4.14}$$

Taking into account (4.9), the natural boundary conditions can be rewritten as follows:

$$\begin{array}{c} r[(a-b)\cos\chi + 2L\cos\beta]\frac{\partial\lambda}{\partial s} = r[(a-b)\sin\chi + 2L\sin\beta]\frac{\partial q}{\partial s} \\ r[(a-b)\sin\chi - 2L\sin\beta]\frac{\partial\lambda}{\partial s} = r[-(a-b)\cos\chi + 2L\cos\beta]\frac{\partial q}{\partial s} \end{array}. \tag{4.15}$$

Dividing the first equation of the system (4.15) by the second, we obtain:

$$(a-b)^2 = 4L^2, \tag{4.16}$$

or, taking into account (4.1),

$$a - 1 = V_{\mathrm{a}} = |L|. \tag{4.17}$$

If $L \geq 0$, then

$$V_{\mathrm{a}} = -V_{\mathrm{b}} = L. \tag{4.18}$$

If $L \leq 0$, then

$$V_{\mathrm{a}} = -V_{\mathrm{b}} = -L. \tag{4.19}$$

If Eq. (4.18) is satisfied at the boundary, then the first equation of system (4.15) can be rewritten as follows:

$$\frac{\sin\beta + \sin\chi}{\cos\beta + \cos\chi} = \frac{\frac{\partial\lambda}{\partial s}}{\frac{\partial q}{\partial s}}, \tag{4.20}$$

or,

$$\tan\frac{\beta + \chi}{2} = \tan\alpha_{\mathrm{s}}, \tag{4.21}$$

or, taking into account (2.13),

$$\alpha_0 = \alpha_s + \pi j, \quad (j = 0,\ 1) \tag{4.22}$$

If Eq. (4.19) is satisfied at the boundary, then the solution is similar.

The above formulae prove the following theorem (Novikova 1986, 1989):

The best projection according to the Airy criterion in the set of close-to-equal-area projections is a projection in which $\ln r(1 + L)$ and $(-\beta)$ are conjugate harmonic functions. At the boundary of the region, the maximum linear distortion equals the function $|L|$, the minimum linear distortion equals the function $-|L|$, and

Tissot's indicatrixes are oriented along tangents or normals to the boundary of the region.

If the linear distortions are quantities of the first order of smallness, then to find the best projection approximately, one can use the requirement that the function $\ln(1+L)$ deviates least from zero in the mapping area. Because:

$$\Delta \ln r(1+L) = \Delta \ln(1+L) + \Delta \ln r = 0, \tag{4.23}$$

if the function $\Delta \ln r$ preserves the sign in the mapping area, the function $\Delta \ln(1+L)$ is either subharmonic or superharmonic. For a superharmonic or subharmonic function to deviate least from zero, it must satisfy Grave's lemma (Graves 1911), i.e., at the boundary, the following equality must be satisfied:

$$\ln(1+L) \approx H = \text{const}. \tag{4.24}$$

If $\Delta \ln\ r \leq 0$, then H is the positive maximum of the function $\ln(1+L)$, and the minimum equal to $(-H)$ is reached by the function $\ln(1+L)$ at some point inside the region.

If $\Delta \ln\ r \geq 0$, then H is the negative minimum of the function $\ln(1+L)$, and the maximum equal to $(-H)$ is reached by the function $\ln(1+L)$ at some point inside the region.

4.3 Relationships Between Mapping Functions and Green's Function

Using (4.10), we obtain:

$$\begin{aligned} \Delta N &= \tfrac{\partial}{\partial q}(2r\cos\beta) + \tfrac{\partial}{\partial \lambda}(2r\sin\beta) \\ \Delta E &= \tfrac{\partial}{\partial \lambda}(2r\cos\beta) - \tfrac{\partial}{\partial q}(2r\sin\beta) \end{aligned}, \tag{4.25}$$

or, taking into account (3.23, 3.25, 3.26):

$$\begin{aligned} N(A) &= 2\iint r|\nabla G(A)|\cos\psi \,\mathrm{d}q\mathrm{d}\lambda + \overline{N}(A) \\ E(A) &= 2\iint r|\nabla G(A)|\sin\psi \,\mathrm{d}q\mathrm{d}\lambda + \overline{E}(A) \end{aligned}, \tag{4.26}$$

where $\overline{N}$, $\overline{E}$ are harmonic functions coinciding on the boundary with N, E and where $\nabla G(A)$ is the gradient of the Green's function (Timan and Trofimov 1968).

4.4 Properties of the Best Projection

Let us show that if the functions $\ln(1+L)$ and V_a reach an extremum at one point inside the mapping area, then at this point, there are no distortions of the angles, i.e.,

$$a = b. \tag{4.27}$$

Indeed, at the extremum point of the function $\ln(1+L)$, the following equalities must be satisfied:

$$(\ln(1+L))_q = (\ln(1+L))_\lambda = 0. \tag{4.28}$$

Taking into account (4.28), the equalities (4.13) take the form:

$$\tau_q = -\beta_\lambda, \quad \tau_\lambda = \beta_q. \tag{4.29}$$

Since at this point, both functions $a,\ b$ have an extremum, then:

$$a_q = a_\lambda = b_q = b_\lambda = 0. \tag{4.30}$$

In this case, the Euler–Urmaev system (2.11, 2.12) is written as follows:

$$\begin{array}{l}(a-b)\left(\tau_\lambda - \chi_q\right)\cos 2\alpha_0 - (a-b)\left(\tau_q + \chi_\lambda\right)\sin 2\alpha_0 = 0 \\ (a-b)\left(\tau_\lambda - \chi_q\right)\sin 2\alpha_0 + (a-b)\left(\tau_q + \chi_\lambda\right)\cos 2\alpha_0 = 0\end{array}. \tag{4.31}$$

Thus, considering Eq. (4.1):

$$a = b = 1, \tag{4.32}$$

valid at the extremum point of the functions $\ln(1+L)$ and V_a.

Therefore, at the central point inside the mapping area, all distortions are zero. Distortions of lengths, angles, and areas reach their extreme values at the boundary of the region.

4.5 Relationships Between the Best Close-to-Equal-Area Projection, the Chebyshev Conformal Projection, and the Airy Ideal Projection

If the mapping area is such that the linear distortions in the best projections are quantities of the first order of smallness, then we can conclude that the Airy ideal projection, the best conformal projection, and the best of the set of close-to-equal-area projections are related to each other. Indeed, to find the ideal projection, it is necessary to determine the harmonic function $\ln r(a+b-1)$, which is equal to $\ln r+H$ at the boundary, and then look for the conjugate harmonic function $(-\beta)$. To find the best projection from a set of close-to-equal-area projections, it is necessary to find the harmonic function $\ln r(1+L)$, which is equal to $\ln r+H$ on the boundary. Next, the harmonic function $(-\beta)$ conjugated to $\ln r(1+L)$ is determined. To find the best conformal projection, according to Chebyshev's theorem (Gravé 1911), in the first stage, one should determine the harmonic function $\ln(ra)$, which is equal to $\ln r+H$ at the boundary of the region. Next, the conjugate harmonic function $(-\gamma)$ is determined. Let us denote the harmonic function equal to $\ln r+H$ at the boundary of the region byΦ. Considering that the angleβ, in conformal projections, is equal to:

$$\beta=\arctan\left(\frac{N_\lambda-E_\mathrm{q}}{N_\mathrm{q}+E_\lambda}\right)=\arctan\left(-\frac{2E_\mathrm{q}}{2N_\mathrm{q}}\right)=\gamma, \tag{4.33}$$

it can be stated that angle β is the same harmonic function used in finding the three projections under consideration.

Let us denote the mapping functions of the Airy ideal projection by $N_1,\ E_1$ and represent them as the sum of two functions:

$$N_1=\overline{N}+u,\quad E_1=\overline{E}+v, \tag{4.34}$$

where $\overline{N},\overline{E}$ are conjugate harmonic functions, the first derivatives of which satisfy the relations:

$$\begin{aligned}\overline{N}_q&=r(a+b-1)\cos\beta=\exp\Phi\cos\beta\\ \overline{N}_\lambda&=r(a+b-1)\sin\beta=\exp\Phi\sin\beta\\ \overline{E}_q&=-r(a+b-1)\sin\beta=-\exp\Phi\sin\beta\\ \overline{E}_\lambda&=r(a+b-1)\cos\beta=\exp\Phi\cos\beta\end{aligned}. \tag{4.35}$$

Since the function $(a+b-1)$ of the ideal projection is equal to the linear scale factor a of the best Chebyshev conformal projection, taking into account formulae (2.9, 2.10), we can conclude that $\overline{N},\ \overline{E}$ are the mapping functions of the best Chebyshev projection.

In reference to Eq. (3.20), the Laplace operators for the functions $u,\ v$ can be determined as follows:

$$\begin{aligned}\Delta u&=\tfrac{\partial}{\partial q}(r\cos\beta)+\tfrac{\partial}{\partial\lambda}(r\sin\beta)\\ \Delta v&=\tfrac{\partial}{\partial\lambda}(r\cos\beta)-\tfrac{\partial}{\partial q}(r\sin\beta)\end{aligned}. \tag{4.36}$$

Let us determine the derivatives of the functions $u,\ v$ with respect to q and λ at the boundary of the region.

$$\begin{aligned} u_{\mathrm{q}} &= N_{1\,\mathrm{q}} - \overline{N}_{q}, \quad u_{\lambda} = N_{1\lambda} - \overline{N}_{\lambda} \\ v_{\mathrm{q}} &= E_{1\,\mathrm{q}} - \overline{E}_{q}, \quad v_{\lambda} = E_{1\lambda} - \overline{E}_{\lambda} \end{aligned}, \tag{4.37}$$

or considering Eqs. (2.9, 2.10, 4.35):

$$\begin{aligned} u_{\mathrm{q}} &= \tfrac{r}{2}[-(a+b-2)\cos\beta + (a-b)\cos\chi] \\ u_{\lambda} &= \tfrac{r}{2}[-(a+b-2)\sin\beta + (a-b)\sin\chi] \\ v_{\mathrm{q}} &= \tfrac{r}{2}[(a+b-2)\sin\beta + (a-b)\sin\chi] \\ v_{\lambda} &= \tfrac{r}{2}[-(a+b-2)\cos\beta - (a-b)\cos\chi] \end{aligned}. \tag{4.38}$$

Since condition (3.12) is satisfied at the boundary, then:

$$\begin{aligned} u_{\mathrm{q}} &= r(a-b)\sin\tfrac{\beta+\chi}{2}\sin\tfrac{\beta-\chi}{2} \\ u_{\lambda} &= -r(a-b)\cos\tfrac{\beta+\chi}{2}\sin\tfrac{\beta-\chi}{2} \\ v_{\mathrm{q}} &= r(a-b)\sin\tfrac{\beta+\chi}{2}\cos\tfrac{\beta-\chi}{2} \\ v_{\lambda} &= r(a-b)\cos\tfrac{\beta+\chi}{2}\cos\tfrac{\beta-\chi}{2} \end{aligned}. \tag{4.39}$$

Let us find the derivatives of the functions $u,\ v$ along the normal and tangent to the boundary of the region. According to (Bronstejn et al. 2020), the normal and tangent derivatives of the function u are equal:

$$u_n = u_{\mathrm{q}}\sin\alpha_s - u_{\lambda}\cos\alpha_s,\ u_s = u_{\mathrm{q}}\cos\alpha_s + u_{\lambda}\sin\alpha_s. \tag{4.40}$$

Similar formulae can be written for the function v.

Using formulae (2.13) and (3.19), we obtain:

$$\begin{aligned} u_n &= -r(a-b)\sin A_0, \quad u_{\mathrm{s}} = 0 \\ v_n &= r(a-b)\cos A_0, \quad v_{\mathrm{s}} = 0 \end{aligned}. \tag{4.41}$$

Since at the border:

$$a + b - 2 = a - b = \exp H - 1, \tag{4.42}$$

then:

$$\begin{aligned} u_n &= -r(\exp H - 1)\sin A_0, \quad u_{\mathrm{s}} = 0 \\ v_n &= r(\exp H - 1)\cos A_0, \quad v_{\mathrm{s}} = 0 \end{aligned}. \tag{4.43}$$

Let us denote the mapping functions of the best projection from the set of close-to-equal-area projections by $N_2,\ E_2$ and represent them as the sum of two functions:

$$N_2 = \overline{N} + u', \quad E_2 = \overline{E} + v', \tag{4.44}$$

where $\overline{N}, \overline{E}$ are conjugate harmonic functions, the first derivatives of which satisfy the relations:

$$\begin{aligned} \overline{N}_q &= r(1+L)\cos\beta = \exp\Phi\cos\beta \\ \overline{N}_\lambda &= r(1+L)\sin\beta = \exp\Phi\sin\beta \\ \overline{E}_q &= -r(1+L)\sin\beta = -\exp\Phi\sin\beta \\ \overline{E}_\lambda &= r(1+L)\cos\beta = \exp\Phi\cos\beta \end{aligned}. \tag{4.45}$$

As in the previous case, $\overline{N}, \overline{E}$ are the mapping functions of the best Chebyshev conformal projection.

Taking into account formulae (4.25), we find the Laplace operators of the functions u', v'.

$$\begin{aligned} \Delta u' &= 2\left[\tfrac{\partial}{\partial q}(r\cos\beta) + \tfrac{\partial}{\partial\lambda}(r\sin\beta)\right] \\ \Delta v' &= 2\left[\tfrac{\partial}{\partial\lambda}(r\cos\beta) - \tfrac{\partial}{\partial q}(r\sin\beta)\right] \end{aligned}. \tag{4.46}$$

Let us find the derivatives of the functions u', v' with respect to q and λ at the boundary of the region:

$$\begin{aligned} u'_{\mathrm{q}} &= N_{2\,\mathrm{q}} - \overline{N}_{\mathrm{q}}, u'_\lambda = N_{2\lambda} - \overline{N}_\lambda \\ v'_{\mathrm{q}} &= E_{2\,\mathrm{q}} - \overline{E}_{\mathrm{q}}, v'_\lambda = E_{2\lambda} - \overline{E}_\lambda \end{aligned}, \tag{4.47}$$

or, taking into account (2.9, 2.10, 4.35):

$$\begin{aligned} u'_q &= 2rL\sin\tfrac{\beta+}{2}\sin\tfrac{\beta-}{2} \\ u' &= -2rL\cos\tfrac{\beta+}{2}\sin\tfrac{\beta-}{2} \\ v'_q &= 2rL\sin\tfrac{\beta+}{2}\cos\tfrac{\beta-}{2} \\ v' &= -2rL\cos\tfrac{\beta+}{2}\cos\tfrac{\beta-}{2} \end{aligned}. \tag{4.48}$$

Because on the border:

$$L = \exp H - 1, \tag{4.49}$$

then:

$$\begin{aligned} u'_n &= 2r(\exp H - 1)\sin \mathrm{A}_0, u'_s = 0 \\ v'_n &= 2r(\exp H - 1)\cos \mathrm{A}_0, v'_s = 0 \end{aligned} \tag{4.50}$$

Let us introduce the notation:

$$u'' = 2u - u', v'' = 2v - v', \tag{4.51}$$

Taking into account (4.36, 4.43, 4.46, 4.50), we obtain:

$$\Delta u'' = 0, \Delta v'' = 0, \tag{4.52}$$

$$u''_n = 0, u''_s = 0, v''_n = 0, v''_s = 0. \tag{4.53}$$

Analyzing formulae (4.52, 4.53), we conclude that u'', v'' are some constants. Since these constants correspond to the choice of the origin of the coordinate system when constructing projections, they can be chosen arbitrarily, for example, set to zero. Then:

$$u' = 2u, \ v' = 2v, \tag{4.54}$$

The derivatives along the tangent to the boundary of the mapping region for the functions u, v, u', v' are equal to zero. This means that these functions are constants at the boundary of the region. These constants, as in the previous case, can be chosen arbitrarily, for example, they can be set to zero. Thus, the mapping functions of the ideal projection, the best conformal one, and the best of the set of close-to-equal-area projection will coincide on the boundary of the region. If we construct the three abovementioned projections in the same coordinate system, then the images of the border of the mapping area will coincide.

Let us show that the images of the extremum point of the subharmonic (superharmonic) function $\Phi - \ln r$ inside the mapping area also coincide.

As shown above, at this point, the function $(a - b)$ in the aforementioned projections also reaches its extremum and is equal to zero. Let us denote this point as A_0. Table 4.1 provides the values of distortions at point A_0 for all of the above projections. It also shows extremal linear distortions of projections that are achieved at the boundary of the region.

Table 4.1 Values of distortions at the extremum point of the function $\Phi - \ln r$ and at the boundary of the region on the ideal, best conformal and best close-to-equal-area projections

Distortion	Projection		
	Best conformal	Airy ideal	Best of the set of close-to-equal-area
At the extremum point of the function $\Phi - \ln r$			
V_p	$-2H$	$-H$	0
V_a	$-H$	$-H/2$	0
V_b	$-H$	$-H/2$	0
ω	0	0	0
On the border			
V_p	$2H$	H	$-H^2 \approx 0$
V_a	H	H	H
V_b	H	0	$-H$
ω	0	H	$2H$

Let the region O of the surface with boundary Γ be mapped with the ideal, best conformal, and best of the set of close-to-equal-area projections onto the region O′ of the plane with boundaryΓ′. Consider a set of inverse projections mapping the region O′ of the plane with boundary Γ′ onto the region O of the surface with boundary Γ. Among this set of projections, there is, obviously, an ideal projection, the best conformal projection, and the best projection from the set of close-to-equal-area projections, and these projections must satisfy the corresponding theorems on the best projections. Moreover, they must have all the properties of the above projections; in particular, there must be a point A_0' on the plane at which all the distortions of the inverse projections reach their extremum. However, since the inverse projections, i.e., ideal, best conformal, and best of the set of close-to-equal-area projections, are projections inverse to the corresponding best direct projections, the point A_0' is the image of the point A_0 of the surface. Consequently, when images of the region O in the three abovementioned projections are constructed in one coordinate system, not only the images of the boundary of the region coincide, but also the images of the point where all the distortions of the projections are extremal.

Taking into account formulae (4.34), (4.44) and (4.54), we can write:

$$N_j = \overline{N} + j \cdot u,\, E_j = \overline{E} + j \cdot v. \tag{4.55}$$

When $j = 0$, the formulae yield the mapping functions of the best Chebyshev conformal projection; when $j = 1$, the mapping functions correspond to the ideal projection according to the Airy criterion; and when $j = 2$, the mapping functions of the best projection from the set of close-to-equal-area projections are used.

Obviously, the mapping functions of an ideal projection are equal to the arithmetic mean of the mapping functions of the best conformal projection and the best of the set of close-to-equal-area projection:

$$N_1 = \frac{1}{2}\left(\overline{N} + N_2\right), \quad E_1 = \frac{1}{2}\left(\overline{E} + E_2\right). \tag{4.56}$$

Let an arbitrary point A'' in the region O be represented in the best conformal projection by the pointA_0'', in the ideal projection by the pointA_1'', and in the best projection from the set of close-to-equal-area projections by the pointA_2''. Taking into account (4.56), we can conclude that if all three projections are constructed in the same coordinate system, then the points, A_0'', A_1'', A_2'', will lie on the same straight line, and A_1'' positioned in the middle between the points A_0'', A_2''.

Thus, three projections, best conformal, ideal according to the Airy criterion, and best from a set of close-to-equal-area projections, form a **triad** of interconnected projections.

All projections from the set of close-to-equal-area projections lose their property of being close-to-equal-area over large areas. Therefore, the best projection considered here is interesting for regions with linear distortions that are quantities of the first order of smallness. An example of such a projection for an area bounded by an almucantar, i.e., among azimuthal projections, is presented in Appendix C.

References

Airy GB (1861) Explanation of a projection by balance of errors for maps applying to a very large extent of the Earth's surface and comparison of this projection with other projections. London, Edinburgh, and Dublin Philosophical Magazine 4 22(149):409–421. https://doi.org/10.1080/14786446108643179

Alekseyev VM, Tikhomirov VM, Fomin SV (1979) Optimal'noye upravleniye (The optimal control). Nauka, Moskow, p 432

Bronstejn IN, Semendyaev KA, Nickel H, Marsolek L, Brunner J et al (2020) *Taschenbuch der Mathematik*. 11, aktualisierte Auflage. Haan-Gruiten, p 1280. ISBN 978-3-8085-5792-1

Canters F, Crols T (2011) Low-error equal-area map projections for global mapping of the terrestrial environment. Vrije Universiteit Brussel, Brussel, Belgium CO-302, p 8

Cheng Y, Lorre JJ (2000) Equal area map projection for irregularly shaped objects. Cartogr Geogr Inf Sci 27(2):91–100. https://doi.org/10.1559/152304000783547957

Deakin RE (1990) A minimum-error equal-area pseudocylindrical map projection. Cartogr Geogr Inf Syst 17(2):161–167. https://www.mygeodesy.id.au/documents/Minimum%20Error%20Deakin%201990.pdf

Gravé D (1911) Demonstration d'un theoreme de Tchebychef generalize. Journal fur die reine und angewandte Mathematik. Band 140. Heft 4 34:247–251. http://resolver.sub.uni-goettingen.de/purl?PPN243919689_0140

Hojovec V (1984) On the general theory of equivalent projection. Studia geophysica et geodaetica 28(4):350–359. https://doi.org/10.1007/BF01642988

Kavraysky VV (1933) Issledovaniya po matematicheskoy kartografii (Studies in mathematical cartography). Trudy TSNIIGAiK, v 6, Leningrad, p 150. https://gis-lab.info/docs/tsniigaik-trudy/tsniigaik-trudy-006-1933.pdf

Kavraysky VV (1934) Matematicheskaya kartografiya (Mathematical cartography). Editorial Department of the ALLUNION Cartographic Trust. Moskow-Leningrad, p 276. https://elib.rgo.ru/handle/123456789/218456

Korn GA, Korn TM (1968) Mathematical handbook: for scientists and engineers. McGraw-Hill Book Company. New York, San Francisco, Toronto, London, Sydney, p 832. https://archive.org/details/mathematicalhand0000gran/page/n7/mode/2up

Novikova EN (1986) Problema otyskaniya nailuchshikh proyektsiy v semeystve blizkikh k ravnovelikim (The problem of finding the best projections in a set of close to equal areas). Krivoy Rog, p 21. Bibliogr. 20 nazv. Rus. Dep. V UkrNIINTI, 22.09.1986, N 2071 Uk-86Dep

Novikova EN (1989): Ideal'naya po kriteriyu Eyri i nailuchshaya iz semeystva blizkikh k ravnovelikim proyektsii dlya obshirnykh oblastey proizvol'nykh ochertaniy (Ideal according to the Airy criterion and the best of the family of close-to- equal area projections for large areas of arbitrary shape). Krivoy Rog, p 53. Bibliogr. 14 nazv. Rus. Dep. v UkrNIINTI, 8.06.1989, N 1567 Uk-89Dep

Peters AB (1984) Distance-Related Maps. Am Cartogr 11(2):119–131. https://doi.org/10.1559/152304084783914821

Snyder JP (1988) New equal-area map projections for noncirculare regions. Am Cartogr 15(4):341–355. https://doi.org/10.1559/152304088783886784

Timan AF, Trofimov VN (1968) Vvedeniye v teoriyu garmonicheskikh funktsiy (Introduction to the theory of harmonic functions). Nedra, Moskow, p 208

Tsinger NA (1898) Kurs vysshey geodezii (Course of higher geodesy). St. Petersburg, p 223

Tsinger NA (1916) O naivygodneyshikh vidakh konicheskikh proyektsiy (On the most advantageous types of conic projections). Izvestiya AN, 6-ya seriya 10(17):1692–1704

Chapter 5
Airy Projection

5.1 Airy Projection Formulae

The extremal linear scale factors of azimuthal projections of the sphere onto a plane in the polar coordinate system $z, \ \alpha$ can be written in the following form (Idrizi 2010, p. 110):

$$\mu_1 = \frac{\rho_z}{R}, \mu_2 = \frac{\rho}{R \sin z}. \tag{5.1}$$

where R is the radius of the sphere, z is the zenith distance of a point on the sphere, and ρ_z is the derivative of the function ρ with respect to z.

Let us introduce auxiliary isometric coordinates $q', \ \lambda'$, associated with geodetic polar coordinates $z, \ \alpha$, by the following formulae:

$$\lambda' = \alpha, \quad q' = \ln\left(\tan\frac{z}{2}\right). \tag{5.2}$$

If the pole of the geodetic system coincides with the pole of the geographic coordinate system, then:

$$\lambda' = -\lambda, \quad q' = -q. \tag{5.3}$$

The minus sign in the first formula of equalities (5.3) arises because the azimuth α is measured clockwise from the initial direction, whereas the longitude λ is counted counterclockwise from the prime meridian. If the initial direction from which the azimuth is measured coincides with the prime meridian, then the directions of measurement for angles α and λ will be opposite.

In direct projections, the relationship between z and φ is determined by the following condition:

E. Novikova, *Best Map Projections*, https://doi.org/10.1007/978-3-031-78334-0_5

$$z = 90^\circ - \varphi.$$

When transitioning from a polar coordinate system to a geodetic system in formulae, the following condition is used:

$$z \to 90^\circ - \varphi. \tag{5.4}$$

The isometric latitude q' is selected in such a way that an increase in this function coincides with an increase in the zenith distance z and not in latitude q. This is why the minus sign is chosen in the second formula of equalities (5.3). To transition from isometric latitude to geodetic latitude on the surface of the sphere, the well-known Eq. (1.11) is used, which, taking into account (5.3, 5.4), is written as follows:

$$q = -q' = \ln\left(\tan\left(90 - \frac{z}{2}\right)\right) = \ln\cot\frac{z}{2} = -\ln\tan\frac{z}{2}. \tag{5.5}$$

Using formulae (5.1, 5.2), we write the Airy criterion for this problem in the following form (Grafarend and Krumm 2006, p. 86):

$$\Xi_A^2 = \frac{1}{2\sigma}\iint\left[\left(\frac{\rho_z}{R} - 1\right)^2 + \left(\frac{\rho}{R\sin z} - 1\right)^2\right] r^2 \mathrm{d}q'\mathrm{d}\lambda'. \tag{5.6}$$

The coefficient of the metric shape of the surface of the coordinate system z, α for the sphere, according to (1.167), is equal to:

$$r = R\sin z. \tag{5.7}$$

Taking into account formulae (5.2) and (5.7), we rewrite Eq. (5.6) as follows (Airy 1861; Lapaine 2016, p. 338):

$$\Xi_A^2 = \frac{1}{2\sigma}\int_0^{2\pi}\int_0^{z_0}\left[\left(\frac{\rho_z}{R} - 1\right)^2 + \left(\frac{\rho}{R\sin z} - 1\right)^2\right] R^2 \sin z \mathrm{d}z\mathrm{d}\alpha. \tag{5.8}$$

Since the integrand in formula (5.8) does not depend on α, then:

$$\Xi_A^2 = \frac{\pi R^2}{\sigma}\int_0^{z_0}\left[\left(\frac{\rho_z}{R} - 1\right)^2 + \left(\frac{\rho}{R\sin z} - 1\right)^2\right] R^2 \sin z \mathrm{d}z. \tag{5.9}$$

Let us introduce the notation:

$$v = \tan\frac{z}{2}. \tag{5.10}$$

Differentiating formula (5.10), we obtain:

$$\frac{dz}{d} = \frac{2}{1+v^2}. \tag{5.11}$$

Using expressions (5.10, 5.11), we replace the variable z in the integral (5.9) with the variable v:

$$\Xi_A^2 = \frac{4\pi R^2}{\sigma} \int_0^{v_0} \left[\left(\frac{\rho_v (1+^2)}{2R} - 1 \right)^2 + \left(\frac{\rho (1+^2)}{2R} - 1 \right)^2 \right] \frac{v}{(1+v^2)^2} dv. \tag{5.12}$$

Therefore, to find the Airy projection, it is necessary to find the minimum of the functional (5.12). The formulated problem is a classical problem of the calculus of variations. As is known (Korn and Korn 1968, p. 350), if the functional Ξ^2 reaches a minimum on some function, ρ, then this function satisfies the Euler differential equation:

$$\frac{\partial Q}{\partial \rho} - \frac{\partial}{\partial v}\left(\frac{\partial Q}{\partial \rho_v} \right) = 0, \tag{5.13}$$

and boundary condition:

$$\left(\frac{\partial Q}{\partial \rho_v} \right)_{v=v_0} = 0. \tag{5.14}$$

In formulae (5.13, 5.14), Q is the integrand function of Eq. (5.12).

Let us find the derivatives included in the expressions (5.13, 5.14).

$$\begin{aligned} \frac{\partial Q}{\partial \rho} &= \frac{4R}{(1+v^2)} \left(\frac{\rho(1+v^2)}{2Rv} - 1 \right) \\ \frac{\partial Q}{\partial \rho_v} &= \frac{4R}{(1+v^2)} \left(\frac{\rho_v(1+v^2)}{2R} - 1 \right) \end{aligned}. \tag{5.15}$$

Substituting (5.15) into (5.13), after identical transformations, yields:

$$\rho_{vv}^2 + \rho_v - \rho = -\frac{4R^3}{(1+v^2)^2}. \tag{5.16}$$

The resulting equation is a linear inhomogeneous equation of the second order, the general solution of which can be written in the following form (Kamke 1948, p. 89):

$$\rho = R\left(\frac{1}{v} \ln\left(1+v^2\right) + k_1 v + k_2 \frac{1}{v} \right), \tag{5.17}$$

where k_1, k_2 are the integration constants.

Differentiating Eq. (5.17) with respect to v gives:

$$\rho_v = R\left(\frac{2}{1+v^2} - \frac{1}{v^2}\ln\left(1+v^2\right) + k_1 - k_2\frac{1}{v^2}\right). \tag{5.18}$$

Let us denote the indefinite integral in formula (5.12) as follows:

$$I = \int\left[\left(\frac{\rho_v\left(1+v^2\right)}{2R} - 1\right)^2 + \left(\frac{\rho\left(1+v^2\right)}{2R} - 1\right)^2\right]\frac{v}{(1+v^2)^2}\mathrm{d}v. \tag{5.19}$$

It is obvious that:

$$\Xi_A^2 = \frac{4\pi R^2}{\sigma}\cdot I\Big|_{v=0}^{v=v_0}. \tag{5.20}$$

Substitution (5.17, 5.18) into (5.19), after integration and identity transformations, gives:

$$\begin{aligned} \mathrm{I} = {} & \ln^2\left(1+v^2\right)\left(-\frac{1}{4v^2}\right) - \frac{1}{2\left(1+v^2\right)} + k_1^2\frac{v^2}{4} \\ & - k_1\frac{1}{2}\ln\left(1+v^2\right) - k_2^2\frac{1}{4v^2} - k_2\frac{1}{2v^2}\ln\left(1+v^2\right). \end{aligned} \tag{5.21}$$

5.2 Choice of Airy Projection Integration Constants

Let us set a condition:

$$\rho = |_{v=0} = 0. \tag{5.22}$$

From this equation, we have:

$$k_2 = 0. \tag{5.23}$$

Let us return to the boundary condition (5.14), which, taking into account the second equality of the system (5.15) is written as follows:

$$\left(\frac{\partial Q}{\partial \rho_v}\right)_{v=v_0} = \frac{4R_0}{\left(1+v_0^2\right)}\left(\frac{\rho_{v_0}\left(1+v_0^2\right)}{2R} - 1\right) = 0. \tag{5.24}$$

Equation (5.24) has three roots:

1. $v_0 = 0$, or, according to (5.10), $z_0 = 0$.
2. $v_0 \to \infty$, $z_0 = 180$.

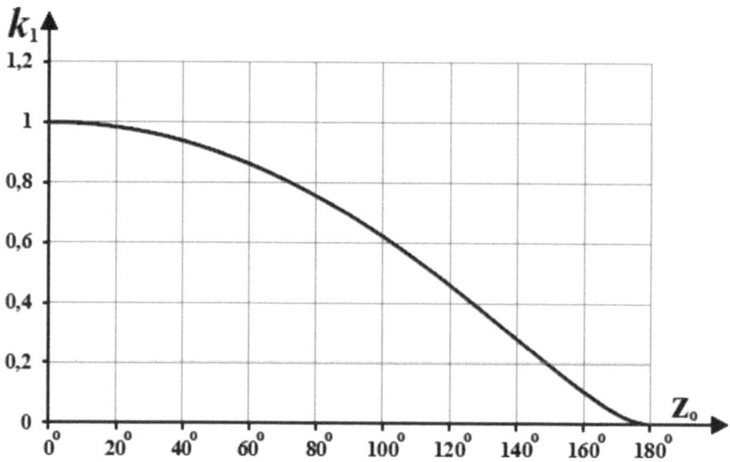

Fig. 5.1 Graph of the change in the constant k_1 with respect to z_0

3. $\frac{\rho_{v_0}(1+v_0^2)}{2R} - 1 = 0.$

The last equality allows us to find the constant k_1. To do this, we substitute formula (5.18) into (5.24). After identical transformations, the following is obtained:

$$k_1 = \frac{1}{v_0^2}\ln(1 + v_0^2). \tag{5.25}$$

Using relation (5.10), the variable v in formulae (5.17) and (5.25) is replaced with the variable z (Lapaine 2016, p. 338).

$$\rho = R\left(2\cot\frac{z}{2}\ln\sec\left(\frac{z}{2}\right) + k_1\tan\frac{z}{2}\right), \tag{5.26}$$

$$k_1 = 2\cot^2\frac{z_0}{2}\ln\sec\left(\frac{z_0}{2}\right). \tag{5.27}$$

Figure 5.1 shows a graph of changes k_1 depending on z_0.

5.3 Theorem for Ideal Projection and Airy Projection

Formulae (5.26, 5.27) completely determine the Airy azimuthal projection. To prove that this projection is Airy ideal for the region bounded by the almucantar, it is sufficient to show that the projection satisfies the ideal projection theorem formulated earlier in Sect. 5.3.

Using equalities (5.1, 5.17, 5.18), we obtain:

$$\mu_1 = \frac{\rho_z}{R} = \frac{\rho_v v_z}{R} = 1 + \frac{\ln \cos \frac{z}{2}}{\sin^2 \frac{z}{2}} + \frac{k_1}{2\cos^2 \frac{z}{2}}, \tag{5.28}$$

$$\mu_2 = -\frac{\ln \cos \frac{z}{2}}{\sin^2 \frac{z}{2}} + \frac{k_1}{2\cos^2 \frac{z}{2}}. \tag{5.29}$$

When $z = z_0$, formula (5.28) is written as follows:

$$\mu_1|_{z=z_0} = 1. \tag{5.30}$$

Taking into account formulae (5.28, 5.29) and (5.7), we find the function $\ln r(a + b - 1)$.

$$\ln(a + b - 1) = \ln\left(k_1 \sec^2 \frac{z}{2}\right), \quad \ln r = \ln(R \sin z) = \ln\left(2R \sin \frac{z}{2} \cos \frac{z}{2}\right), \tag{5.31}$$

or

$$\ln r(a + b - 1) = \ln\left(2Rk_1 \tan \frac{z}{2}\right). \tag{5.32}$$

As is known (Bugayevskiy and Snyder 1995, p. 101), the mapping functions of azimuthal projections are determined by the following formulae:

$$N = \rho \cos \alpha = \rho \cos \lambda', E = \rho \sin \alpha = \rho \sin \lambda'. \tag{5.33}$$

Differentiating formulae (5.33) and taking into account (5.2, 5.26), we obtain:

$$\begin{array}{l} N_{q'} = \rho_{q'} \cos \alpha, N_{\lambda'} = -\rho \sin \alpha \\ E_{q'} = \rho_{q'} \sin \alpha, E_{\lambda'} = \rho \cos \alpha \end{array}. \tag{5.34}$$

The formulae obtained above for finding the ideal projection are invariant with respect to an arbitrary isometric coordinate system. Therefore, the first formula of system (2.6) can be rewritten as follows:

$$\tan \beta = \frac{N_{\lambda'} - E_{q'}}{N_{q'} + E_{\lambda'}}, \tag{5.35}$$

or, taking into account (5.34),

$$\tan \beta = -\tan \alpha. \tag{5.36}$$

Hence,

$$\beta = -\alpha + \pi \cdot j, \ (j = 0, \ 1) \tag{5.37}$$

Let us differentiate the functions $\ln r(a+b-1)$, β with respect to q', λ':

$$\frac{\partial}{\partial q'}(\ln r(a+b-1)) = 1, \quad \frac{\partial}{\partial \lambda'}(\ln r(a+b-1)) = 0, \tag{5.38}$$

$$\frac{\partial \beta}{\partial q'} = 0, \quad \frac{\partial \beta}{\partial \lambda'} = -1. \tag{5.39}$$

Taking into account formulae (2.15, 5.33), we find the azimuth of the main direction of the projection:

$$\tan 2\alpha_0 = 0, \tag{5.40}$$

or,

$$\alpha_0 = \frac{\pi \cdot j}{2}, \quad (j = 0,\ 1). \tag{5.41}$$

Consequently, the main directions are oriented either along the tangent or normal to the lines $q' = \text{const}$.

Formulae (5.37) and (5.41) apply for any azimuthal projection since equalities (5.34) are common for all azimuthal projections.

Analyzing formulae (5.29, 5.38, 5.39, 5.41), we can conclude that the Airy azimuthal projection is ideal according to the Airy criterion for the area limited by the almucantar. Indeed, according to (5.38, 5.39), $\ln r(a+b-1)$ and β are conjugate harmonic functions. At the boundary of the region, the minimum scale factor is equal to unity (equality (5.30)). According to (5.41), the azimuths of the main directions of the projection coincide with the azimuths of the tangents and normals to the boundary of the region.

5.4 Properties of the Airy Projection

Let us show that the Airy projection possesses all the properties of Airy ideal projections.

At the central point for $z = 0$, the functions μ_1, μ_2 are indeterminate since the expression:

$$\frac{\ln \cos \frac{z}{2}}{\sin^2 \frac{z}{2}},$$

is the uncertainty of the form $\frac{0}{0}$. However, since there is a limit:

$$\lim_{z \to 0} \frac{\ln \cos \frac{z}{2}}{\sin^2 \frac{z}{2}} = -\frac{1}{2}, \tag{5.42}$$

then the functions μ_1, μ_2 can be further defined at this point as follows:

$$\mu_1 = \mu_2 = \frac{1}{2} + \frac{k_1}{2}. \tag{5.43}$$

Hence, we can conclude that at the central point of the projection, the extremal linear scale factors are equal, i.e., condition (3.35) is satisfied.

According to the first formula of the system (5.31), the function $\ln(a + b - 1)$ at the central point is equal to:

$$\ln(a + b - 1)|_{z=0} = \ln k_1, \tag{5.44}$$

and at the boundary $\ln(a + b - 1)$ takes the value equal to:

$$\ln(a + b - 1)|_{z=z_0} = \ln\left(k_1 \sec^2 \frac{z_0}{2}\right). \tag{5.45}$$

Let us denote:

$$k_3 = \ln k_1 + \ln\left(k_1 \sec^2\left(\frac{z_0}{2}\right)\right), \tag{5.46}$$

or, with Eq. (5.27),

$$k_3 = 2 \ln\left(2 \cot^2\left(\frac{z_0}{2}\right) \sec\left(\frac{z_0}{2}\right) \ln \sec\left(\frac{z_0}{2}\right)\right). \tag{5.47}$$

Taking into account formula (2.23), we can conclude that for a sphere, as for an ellipsoid of revolution $\Delta \ln r < 0$. Therefore, the function $\ln(a + b - 1)$ is subharmonic. For the subharmonic function $\ln(a + b - 1)$ to deviate from zero the least, the value k_3 must be equal to zero. Figure 5.2 shows a graph of the change in value of k_3 depending on z_0.

Analyzing Fig. 5.2, we can conclude that if the mapping area is such that the distortions of projection lengths are quantities of the first order of smallness, then we can equate k_3 to zero and assume that the subharmonic function satisfies Grave's lemma (Grave 1896, 1911). According to Fig. 5.2, the zenith distance of the boundary almucantar must satisfy the condition $z_0 < 60$.

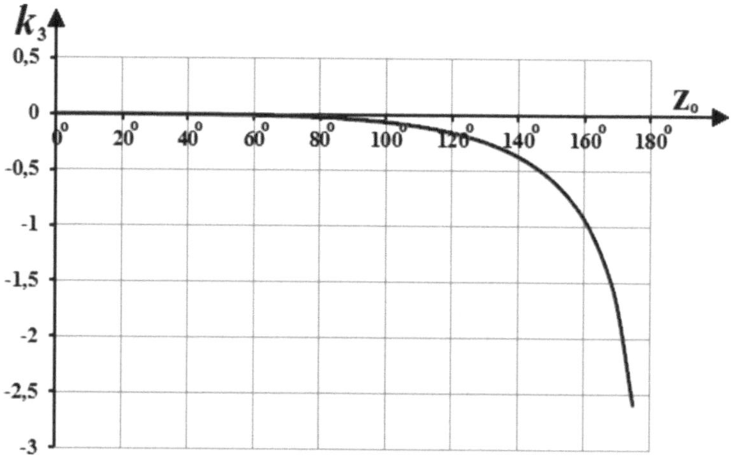

Fig. 5.2 Graph of parameter k_3 changes depending on changes in z_0

5.5 Airy Projection for the Entire Sphere

Theoretically, the zenith distance of the boundary almucantar can be chosen arbitrarily. To apply the projection to the entire sphere, the boundary value of the zenith distance must be equal to 180°. In this case, the entire surface of the Earth is displayed from a zenith distance of 0 to a zenith distance of 180°.

For the Airy projection over the entire territory of the sphere, the formula that determines the constant k_1 for $v_0 \to \infty$ contains an uncertainty of the form $^{\infty}/_{\infty}$. However, this uncertainty has a limit, namely:

$$\lim_{v_0 \to \infty} \frac{\ln(1 + v_0^2)}{v_0^2} = 0. \tag{5.48}$$

Consequently, for the Airy projection over the entire territory of the globe, the constant $k_1 = 0$, which is also clearly shown in Fig. 5.1.

Figure 5.3 shows the map grid of the Airy projection, covering the entire globe, with meridians and parallels drawn every 10 degrees. The polar coordinates of the points at the intersection of meridians and parallels were calculated using the following formulae (Bugayevskiy and Snyder 1995, p. 13):

$$\tan \alpha = \frac{\cos \varphi \sin \lambda}{\sin \varphi}, \quad \cos z = \cos \varphi \cos \lambda. \tag{5.49}$$

The radius of the parallel in the projection, ρ, was calculated using Eq. (5.26) at $k_1 = 0$. Plane rectangular coordinates N, E in the projection were determined using equalities (5.33).

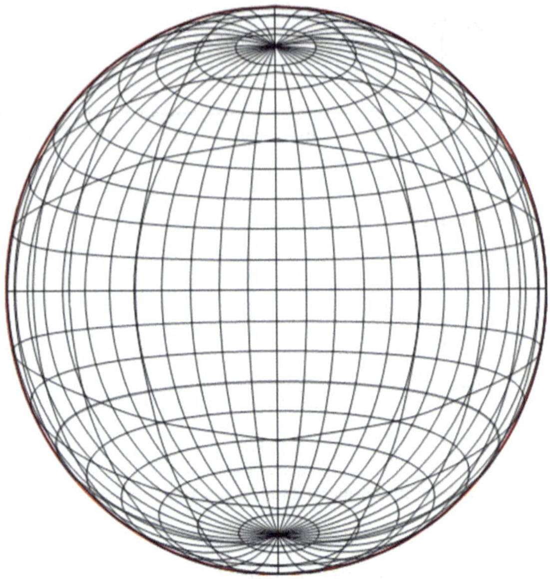

Fig. 5.3 Cartographic grid of the Airy projection, built over the entire territory of the globe

As shown in Fig. 5.3, the Airy projection onto the entire surface of the Earth loses its one-to-one character. It wraps the surface image so that one projection point corresponds to two surface points. Thus, the principle of uniqueness of projections is violated, according to which one point of the projection must correspond to only one point of the original surface.

The value of the Airy projection indicator of unambiguity, according to formulae (1.14, 1.109, 5.2, 5.34), is equal to:

$$\overline{\sigma} = \frac{h}{R^2} = \frac{\rho \cdot \rho_{q'}}{R^2} = \frac{\rho \cdot \rho_v v}{R^2}, \tag{5.50}$$

or, taking into account (5.10, 5.18, 5.26),

$$\overline{\sigma} = -4\cos^2\left(\frac{z}{2}\right)\ln\cos\left(\frac{z}{2}\right)\left(1 + \frac{\ln\cos\left(\frac{z}{2}\right)}{\sin^2\left(\frac{z}{2}\right)}\right). \tag{5.51}$$

Figure 5.4 shows a graph of changes in the indicator of unambiguity for the projection under consideration, constructed according to formula (5.51).

According to the graph of $\overline{\sigma}$, presented in Fig. 5.4, the Airy projection, constructed over the entire territory of the globe, does not satisfy the principle of unambiguity. The interval of change in the zenith distance is divided into two unequal parts. In the first part, when the zenith distance varies within the limits $0 < z < z_{\max}(\rho)$, the indicator is positive; in the second part, with limits $z_{\max}(\rho) < z < 180°$, the

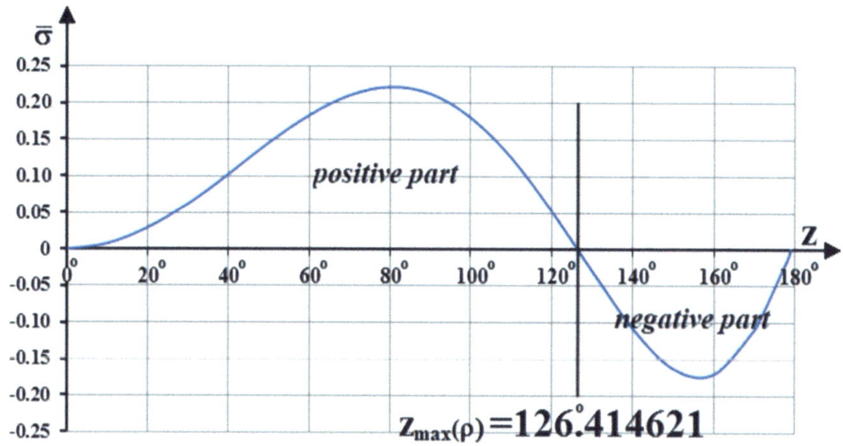

Fig. 5.4 Graph of changes in the indicator of unambiguity depending on the zenith distance z, built for the Airy projection

indicator is negative. At the point where the zenith distance $z_{\max}(\rho) = 126°.414621$, the indicator changes its sign (Fig. 5.5).

To analyze the reason for the nonfulfillment of the principle of unambiguity of the Airy projection, let us refer to the graph of changes in the function ρ, constructed using formulae (5.26, 5.27).

Note that the graph has a maximum for the zenith distance $z_{\max}(\rho) = 126.414621$. It is at this point that the indicator of unambiguity changes its sign. Furthermore, as

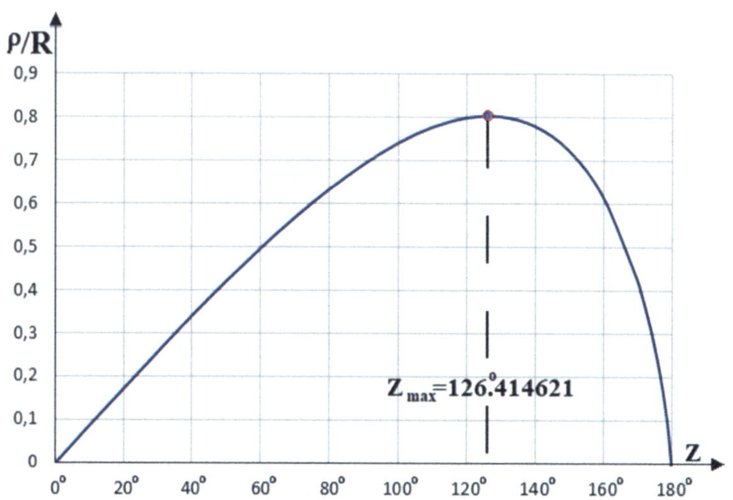

Fig. 5.5 Graph of change ρ depending on z for the Airy azimuthal projection, constructed for the territory of the entire globe in units of radius (for $R = 1$)

the zenith distance increases, ρ decreases; this is why the image of the globe after the zenith distance equals $z_{max}(\rho)$ looks as if it turns on itself, violating the principle of unambiguity. One point in the projection corresponds to two points on the surface of the globe. The entire surface of the globe is divided into two unequal parts. In the first part for $0 < z < z_{max}(\rho)$, the polar distance increases from 0 to 0.804742 (in units of the radius). Conversely, in the second part for $z_{max}(\rho) < z < 180$, the polar distance decreases from 0.804742 to 0. Within each part, the principle of unambiguity is maintained. In Fig. 5.6, the map grids of both parts are shown separately.

Figure 5.6 shows in red the circle corresponding to the almucantar with the maximum value ρ for which $z_{max}(\rho) = 126.414621$. Thus, the limiting value of the zenith distance at which the projection satisfies the principle of unambiguity is equal to $z_{max}(\rho) = 126.414621$.

Figure 5.7 presents graphs of extremal linear scale factors depending on the zenith distance of the Airy projection constructed for the entire globe.

The graphs in Fig. 5.7 do not apply to the value $z = 180$, since both extremal scale factors tend to infinity for this value. Moreover, the point where the scale tends to infinity, i.e., the point with a zenith distance equal to 180°, does not go to infinity but is depicted as a point in the very center of the projection. This occurs because although the area scale factor at this point tends to infinity, the Jacobian at this point is zero, as shown in Fig. 5.4.

According to (5.2, 5.7, 5.10, 5.34, 5.51), the area scale factor of the Airy projection is determined by the following equation:

$$p = -\frac{\ln\cos\frac{z}{2}}{\sin^2\frac{z}{2}}\left(1 + \frac{\ln\cos\frac{z}{2}}{\sin^2\frac{z}{2}}\right). \tag{5.52}$$

As shown in Fig. 5.7, at the boundary of the mapping area (for this area, the boundary is a point with coordinates $z_0 = 180°$), the minimum scale factor is not equal to one. Nevertheless, at this point, the natural boundary conditions are satisfied since $z_0 = 180°$ is one of the roots of Eq. (5.24).

The parameters of the resulting projection are presented in Table 5.1. The value of the Airy criterion is 71%. Despite the fact that the boundary point with $z_0 = 180°$ is depicted in the projection as a point, the maximum distortions of lengths, areas, and angles at this point tend to infinity.

5.6 *Airy Projections Satisfying the Principle of Unambiguity*

Notably, the function ρ has a maximum inside the mapping area, not only when $z_0 = 180°$. It turns out that for each value $\overline{z}_0 \le z_0 \le 180°$, there is a unique value z_{max}, and the numerical value z_{max} depends on the choice $\overline{z}_0$.

The dependence of z_{max} (ν_{max}) on z_0 (ν_0) can be determined by using Eq. (5.18), taking into account that for z_{max}, the derivative ρ_ν is as follows:

Fig. 5.6 Cartographic grid of the Airy projection, built over the entire territory of the globe, divided into regions of unambiguity: **a** the grid in the positive region of unambiguity with zenith distances $0 < z < z_{\max}(\rho)$; **b** the grid in the negative region of uniqueness with zenith distances $z_{\max}(\rho) < z < 180$

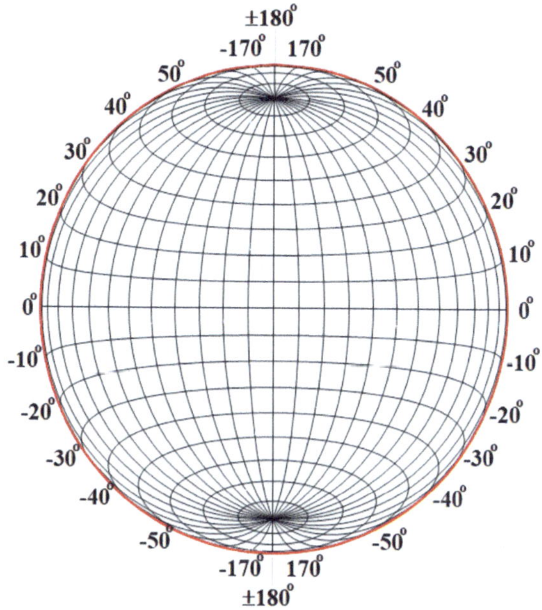

(a)

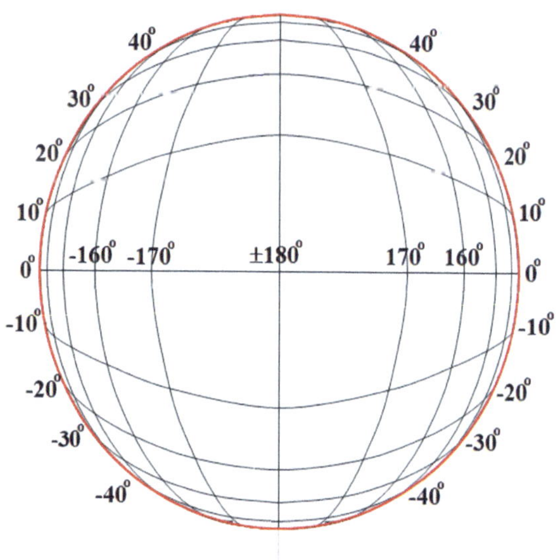

(b)

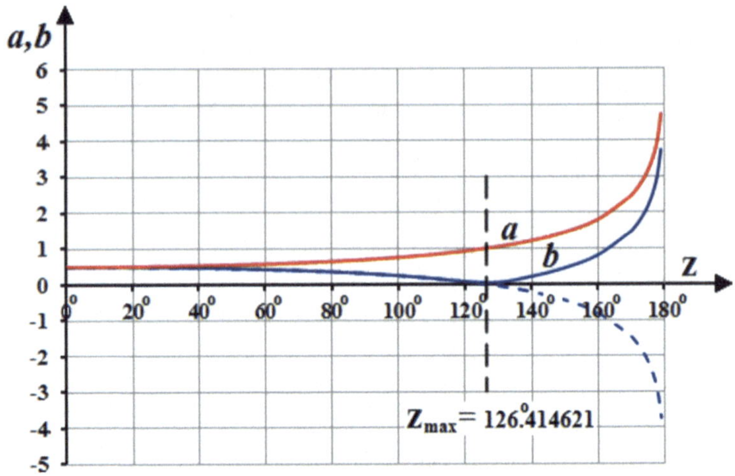

Fig. 5.7 Graphs of extremal linear scale factors of the Airy projection constructed for the entire globe

Table 5.1 Airy projection options parameters

Parameter	Options of Airy projection		
$z_0(°)$	180	166	149.329254
Percentage of globe surface coverage	100	98,5	93
$V_{b\ min}(\%)$	−50	−47	−40
$V_{a\ min}(\%)$	−50	−47	−40
$V_{b\ max}(\%)$	+∞	0	0
$V_{a\ max}(\%)$	+∞	327	186
$V_{p\ min}(\%)$	−75	−72	−64
$V_{p\ max}(\%)$	+∞	327	186
$\omega_{min}(°)$	0	0	0
$\omega_{max}(°)$	+∞	77	58
$\Xi_A(\%)$	71	60	46

$$\rho_v|_{v=v_{max}} = 0, \tag{5.53}$$

or,

$$\frac{2}{1+v_{max}^2} - \frac{1}{v_{max}^2}\ln(1+v_{max}^2) + k_1 = 0. \tag{5.54}$$

Considering (5.27), we find:

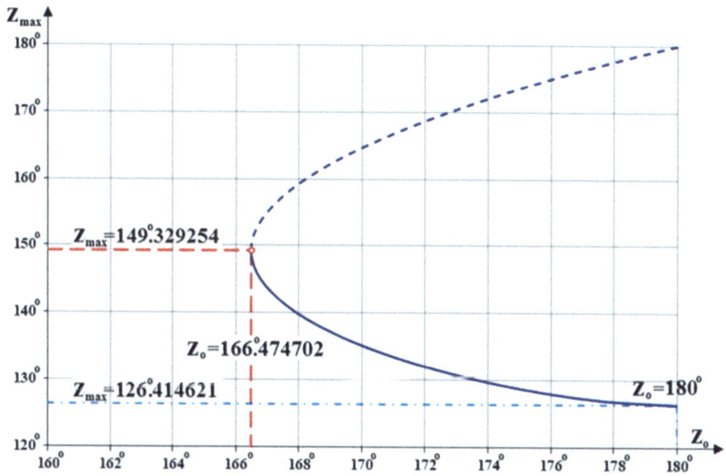

Fig. 5.8 Graph of $z_{\max}$ (z value for which the function ρ reaches a maximum) versus z_0 (zenith distance of the boundary almucantar)

$$\frac{2}{1+v_{\max}^2}-\frac{1}{v_{\max}^2}\ln(1+v_{\max}^2)+\frac{1}{v_0^2}\ln(1+v_0^2)=0. \quad (5.55)$$

Equation (5.55) is transcendental with respect to both $\nu_{\max}$, ν_0. Its solution using Newton's method is depicted in Fig. 5.8. The solution is calculated for the values $\nu_{\max}$ and ν_0. For each value ν_0, the corresponding value $\nu_{\max}$ was determined. Next, from the quantities ν, a transition was made to the quantities z by considering Eq. (5.10).

As shown in Fig. 5.8, each value z_0 in the range from 166°.474702 to 180° corresponds to two values $z_{\max}$. Only for the value $z_0 = 166°.474702$ is there one value $z_{\max} = 149°.329254$, not two. At this limit point, both the first and second derivatives of the function ρ are equal to zero.

Figure 5.9 shows graphs of the changes in ρ and ρ_ν depending on z. Both functions are plotted for the value k_1 corresponding to $z_0 = 166°.474702$. This value z_0 and its corresponding value $z_{\max}$ are shown in the graphs of both functions. According to Fig. 5.9, the function ρ has a local maximum at point $z_{\max} = 149°.329254$. This is why the function ρ_ν at this point is zero. However, at this point, the graph of the function ρ_ν does not intersect but touches the axis z, indicating that this point is also the minimum of the function ρ_ν. Owing to this property, the value of $z_{\max}$ can be found by solving the equation:

$$\rho_{\nu\nu}\big|_{\nu=\nu_{\max}}=0, \quad (5.56)$$

or,

$$\frac{2}{v_{\max}^3}\ln(1+v_{\max}^2)-\frac{2\left(1+3v_{\max}^2\right)}{v_{\max}(1+v_{\max}^2)^2}=0. \quad (5.57)$$

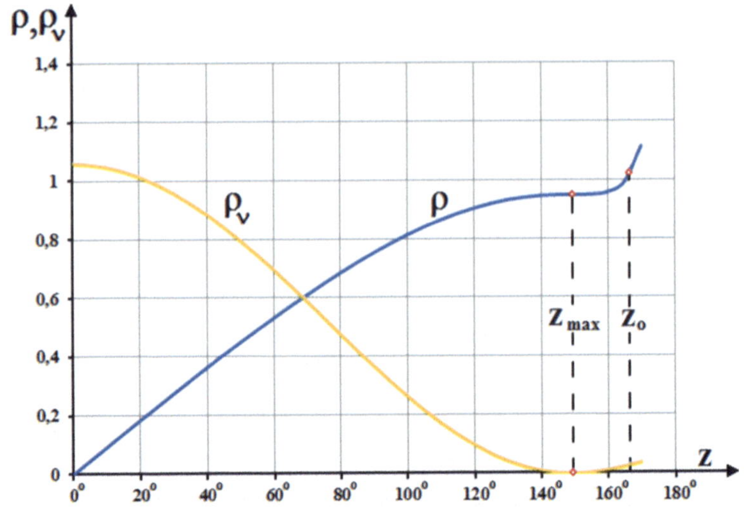

Fig. 5.9 Graphs of changes in ρ and ρ in units of radius (for $R = 1$), depending on z for $z_0 =$ 166°.474702

The transcendental Eq. (5.57) was solved by Newton's method. Solution result:

$$v_{\max} == 3.64653763546304.$$

This value corresponds to:

$$z_{\max} = 149^{\circ}.329254$$

Since $z_0 = 166^{\circ}.474702$ is the limiting value of the quantity z_0, all values of z_0 less than this value, are parameters of the Airy projections, for which the function ρ increases monotonically throughout the mapping area, i.e., in the range from 0 to z_0. This means that both mapping conditions are satisfied for these projections: unambiguity and continuous differentiability.

Figure 5.10 shows the Airy projection map grid with $z_0 = 166°$. Since $166 < 166.474702$, the mapping functions of this projection satisfy both restrictions formulated in Sect. 1.1. The meridians and parallels of the grid are drawn every 30°.

Figure 5.10 also shows Tissot's indicatrixes. The projection covers 98.5% of the entire globe (see Table 5.1). However, unlike the previous version of the projection, which covers 100% of the territory, the distortions of the lengths of lines, areas, and angles on the boundary almucantar do not tend to infinity.

According to Fig. 5.10, the indicatrixes near the boundary almucantar have practically degenerated into segments, indicating large differences between the maximum and minimum linear distortions. To understand this phenomenon, let us return to the

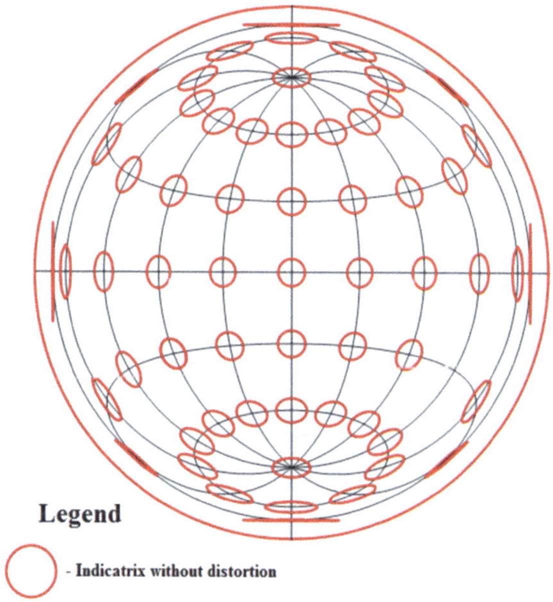

Fig. 5.10 Cartographic grid of Airy projection, built for $z_0 = 166$ with Tissot's indicatrixes. Meridians and parallels are drawn every 30°. The outer circle in red is an image of the border almucantar with $z_0 = 166$

graph of the function ρ (Fig. 5.9). As shown in Fig. 5.9, the function ρ near the value z_{max} practically does not change. This means that the image of the surface of the globe is compressed when it approaches the boundary almucantar. This is why the images of areas close to the boundary almucantar are highly deformed, although they do not overlap.

Figure 5.11a shows a graph of changes in the extremal linear distortions of this version of the Airy projection. Table 5.1 presents its main characteristics. The value of the Airy criterion is 60%, the maximum linear distortion is 327%, and the minimum is −47%.

Therefore, choosing a boundary value close to 166° is not an optimal solution. A more appropriate choice for z_0 is equal to the minimum value z_{max}, i.e., $z_0 = 149°.329254$. A projection with such a boundary almucantar value does not have the disadvantages of the previous version of the projection. Since 149.329254 < 166, the projection mapping functions are also one-to-one and continuously differentiable throughout the mapping region. This projection covers 93% of the entire surface of the globe. Throughout the mapping area, the maximum scale factor does not exceed 186%, and the minimum scale factor does not fall below (−40%) (see Table 5.1). Figure 5.12 shows the cartographic grid of this projection with Tissot's indicatrixes. Figure 5.13 also shows the map projection grid, but with distortions displayed via symbol-charts. The analysis shows that almost everywhere within the mapping area, the linear distortion does not exceed ± 50%, indicating that this projection has the most favorable distortion distribution.

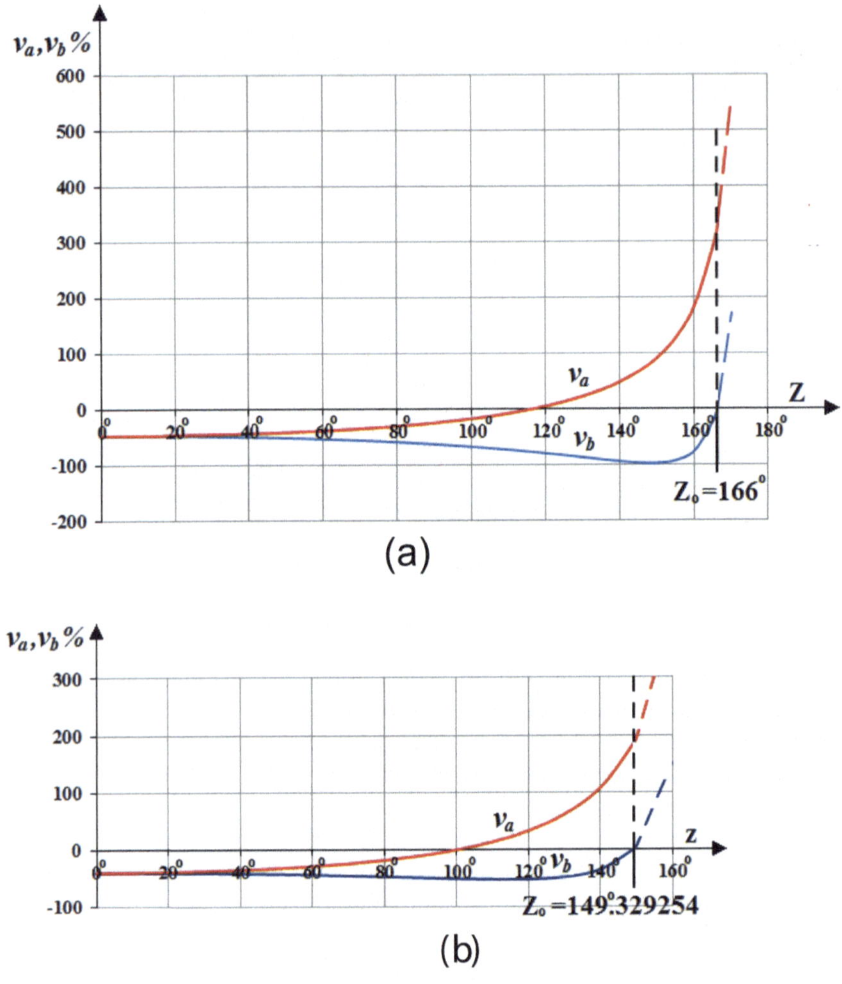

Fig. 5.11 Graphs of changes in extremal linear distortions of Airy projections: **a** for $z_0 = 166°$; **b** for $z_0 = 149°.329254$

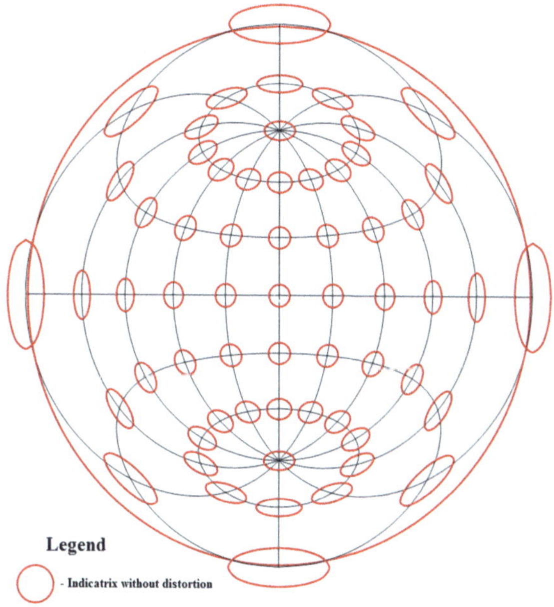

Fig. 5.12 Cartographic grid of the Airy projection, constructed for $z_0 = 149°.329254$ with Tissot's indicatrixes

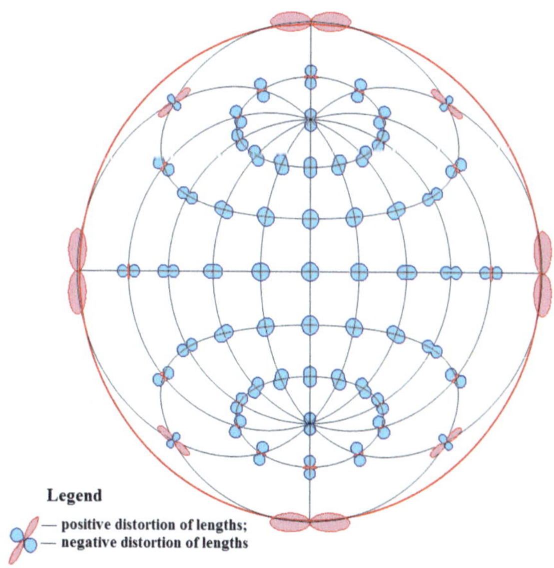

Fig. 5.13 Cartographic grid of the Airy projection, constructed for $z_0 = 149°.329254$ with the symbol-charts

References

Airy GB (1861) Explanation of a projection by balance of errors for maps applying to a very large extent of the Earth's surface and comparison of this projection with other projections. London,

Edinburgh, and Dublin Philosophical Magazine 4, 22(149):409–421. https://doi.org/10.1080/14786446108643179

Bugayevskiy LM, Snyder JP (1995) Map projections. Reference Manual 328 Taylor&Francis Ltd., London. https://doi.org/10.1201/b16431

Grafarend EW, Krumm FW (2006) Map Projections. Cartogr Inf Syst. 713 Springer-Verlag Berlin Heidelberg. https://books.google.com.ua/books/about/Map_Projections.html?id=Rg3Lc8omIXMC&redir_esc=y

Grave DA (1896) Sur la construction des Cartes geographiques Journal de mathematiques pures et appliquees, 5(2):317–361. http://www.numdam.org/item?id=JMPA_1896_5_2__317_0

Gravé D (1911) Demonstration d'un theoreme de Tchebychef generalize. Journal fur die reine und angewandte Mathematik. Band 140. Heft 4, 34:247–251. http://resolver.sub.uni-goettingen.de/purl?PPN243919689_0140

Idrizi B (2010) Hartografia Matematike, Universiteti i Prishtines. Prishtine, p 207. https://www.researchgate.net/publication/312491390_HARTOGRAFIA_MATEMATIKE_-_projeksione_hartografike_Draft_version

Kamke E (1948) Differentialgleichungen Losungsmethoden Und Losungen Band I. Gewohnliche Differentialgleichungen. Chelsea Publishing Company. New York, p 710. https://archive.org/details/in.ernet.dli.2015.362343/page/n1/mode/2up

Korn GA, Korn TM (1968) Mathematical handbook: for scientists and engineers. McGraw-Hill Book Company. New York, San Francisco, Toronto, London, Sydney, p 832. https://archive.org/details/mathematicalhand0000gran/page/n7/mode/2up

Lapaine M (2016) George Biddell Airy and his contribution to map projections theory. Paper presented at the 6th International conference on cartography and GIS. Albena, Bulgaria, pp 13–17. ISSN 1314-0604

Chapter 6
Gauss–Kruger Projection

Before considering the Gauss-Kruger projection (Buchroithner and Pfahlbusch 2016; Deakin and Hunter 2010; Deakin et al. 2012), the properties of its simpler original version, the direct cylindrical projection, i.e., the Mercator projection (Abee 2019; De Jong 1968; Grafarend and Krumm 2006; Daners 2012; Fabregat 2017; Osborn 2013) will be analyzed.

6.1 The Best Airy Projection Among Cylindrical Ones

Let us choose as the mapped area the territory of the Earth's ellipsoid bounded by two parallels with latitudes $\pm\varphi_0$ and corresponding isometric latitudes $\pm q_0$, equally spaced from the equator. Unlike the mapping area discussed in Sect. 5 (Figs. 5.3 and 5.6), the resulting mapped area is not simply connected. This means that there are closed curves that cannot be compressed into one point within the region itself. For example, such curves are parallels inside this area. Figure 6.1 shows two regions: Fig. 6.1a is simply connected, and Fig. 6.1b is doubly connected.

The uniqueness of the mapped area directly affects the properties of the projections used. In particular, projections of a doubly connected region are not limited along the E axis unless forcibly constrained. Notably, if the mapping area is limited not only by two parallels but also by two meridians, even if they coincide on the surface of the Earth's ellipsoid or globe, the resulting mapping area will no longer be doubly connected. This is a simply connected region, the boundaries of which are not only parallels, but also meridians.

Owing to the symmetry of the area under consideration with respect to the coordinate axes N, E and its unboundedness along the E axis, cylindrical projections are used to map it onto the plane. The scale factors along the meridians and parallels for these projections are written as follows (Bugayevskiy and Snyder 1995, p. 49, Ben Hadj Salem and Commiot 2021):

E. Novikova, *Best Map Projections*, https://doi.org/10.1007/978-3-031-78334-0_6

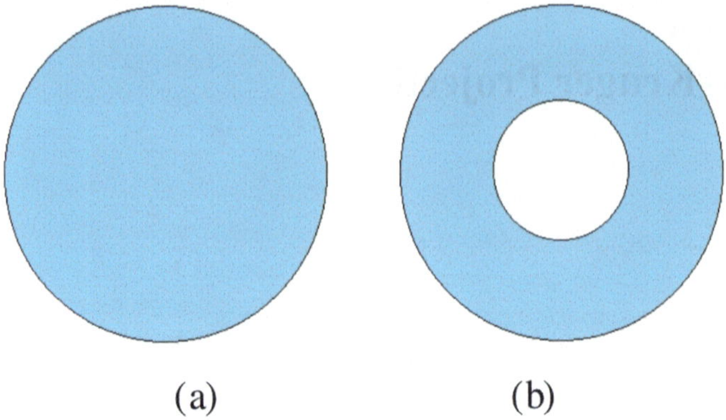

Fig. 6.1 Schematic representation of the mapping area: **a** simply connected; **b** doubly connected

$$m = \frac{N_q}{r}, \quad n = \frac{\kappa}{r}, \tag{6.1}$$

where N_q is the derivative of the mapping function N with respect to the variable q and where κ is the constant value. Moreover, the mapping functions themselves have such limitations (Bugayevskiy and Snyder 1995, p. 49):

$$N = N(q), \quad E = \kappa \cdot \lambda. \tag{6.2}$$

Since the map grid of all cylindrical projections is orthogonal, the extremal scale factors are determined from the relationships (1.142). Therefore, the Airy criterion for the selected mapping area is written in the following form (Grafarend and Krumm 2006, p. 280):

$$\Xi_A^2 = \frac{1}{\sigma} \int_0^{2\pi} \int_{-q_0}^{q_0} \left[\left(\frac{N_q}{r} - 1 \right)^2 + \left(\frac{\kappa}{r} - 1 \right)^2 \right] r^2 \mathrm{d}q \mathrm{d}\lambda. \tag{6.3}$$

As the linear scales of cylindrical projections do not depend on longitude, formula (6.3) can be written as follows:

$$\Xi_A^2 = \frac{2\pi}{\sigma} \int_{-q_0}^{q_0} \left[\left(\frac{N_q}{r} - 1 \right)^2 + \left(\frac{\kappa}{r} - 1 \right)^2 \right] r^2 \mathrm{d}q. \tag{6.4}$$

Obviously, the integrand function of the integral (6.4) is equal to:

$$Q = N_q^2 - 2N_q \cdot r + \kappa^2 - 2\kappa \cdot r + 2r^2. \tag{6.5}$$

For the azimuthal Airy projection (see Sect. 5), the functional Ξ^2 reaches a minimum for the function N if this function satisfies the Euler differential equation and the natural boundary condition (Korn and Korn 1968, p. 350).

For cylindrical projections, the Euler equation and boundary condition are written as follows:

$$\frac{\partial Q}{\partial N} - \frac{\partial}{\partial q}\left(\frac{\partial Q}{\partial N_q}\right) = 0, \tag{6.6}$$

$$\left(\frac{\partial Q}{\partial N_q}\right)_{q=q_0} = 0. \tag{6.7}$$

The substitution of the derivatives of the integrand Q into the Euler equation followed by the integration of the resulting expression gives:

$$N_q - r = C, \tag{6.8}$$

where C is the integration constant. Taking into account the boundary condition:

$$\left(\frac{\partial Q}{\partial N_q}\right)_{q=q_0} = C = 0. \tag{6.9}$$

Integrating Eq. (6.8) under condition (6.9), we obtain:

$$N|_{q=q_0} = \int_0^{q_0} r\mathrm{d}q = \int_0^{\varphi_0} M\,\mathrm{d}\varphi - S. \tag{6.10}$$

In Eq. (6.10), the function S is the length of the meridian arc from the equator to the parallel with latitude φ_0.

Thus, the cylindrical projection, best according to the Airy criterion, is the projection equidistant along the meridians, known since ancient times as the Marinus of the Tyre projection (Snyder 1987, p. 90), and revived in the fifteenth century under the name of the projection of Enrique the Navigator (Alegria et al. 2007). Since this projection is the best of all possible projections according to the Airy criterion for a doubly connected region bounded by two parallels equidistant from the equator, it will henceforth be called the Airy ideal cylindrical projection.

Erik W. Grafarend and Friedrich W. Krumm (2006, p. 279), while comparing three cylindrical projections—conformal, equal-area, and equidistant—showed that the best according to the Airy criterion is the equidistant projection.

Obviously, the value of the Airy criterion in Eq. (6.4) depends not only on the function N but also on the constant κ. Specifically, we perform the integration of Eq. (6.4). Taking into account (6.5):

$$\Xi_A^2 = \frac{2\pi}{\sigma} \int_{-q_0}^{q_0} Q \mathrm{d}q = \frac{2\pi}{\sigma} \left(\kappa^2 2q_0 - 2\kappa \int_{-q_0}^{q_0} r \mathrm{d}q + \int_{-q_0}^{q_0} r^2 \mathrm{d}q \right). \tag{6.11}$$

To find the minimum of this functional, we differentiate Eq. (6.11) with respect to κ and equate the resulting derivative to zero, namely:

$$\frac{\partial}{\partial \kappa} \left(\int_{-q_0}^{q_0} Q \mathrm{d}q \right) = 4\kappa \cdot q_0 - 2 \int_{-q_0}^{q_0} r \mathrm{d}q = 0. \tag{6.12}$$

Hence, taking into account the symmetry of the function $\int_{-q_0}^{q_0} r dq$ with respect to the equator,

$$\kappa = \frac{\int_0^{q_0} r \mathrm{d}q}{q_0}. \tag{6.13}$$

Substituting this value κ into formula (6.11) and performing identical transformations yields:

$$\Xi_A^2 = 1 - \frac{S^2}{P \cdot q_0}, \tag{6.14}$$

Where

$$P = \int_0^{q_0} r^2 \mathrm{d}q. \tag{6.15}$$

Figure 6.2 shows a graph of changes in the value of κ depending on the geodetic latitude φ_0 of the boundary parallel for the case when the WGS84 ellipsoid is used but with a semi-major axis equal to 1. The parameters of this ellipsoid are presented in Appendix D.

The value of κ for the ideal according to Airy projection, constructed for the case when the mapping area is a latitudinal zone with boundary parallels $\varphi_0 = \pm 3°$, is presented in Table 6.1. The values of extremal linear distortions of this projection are also given here. The working formulae and parameters for the calculations of functions S, P are given in Appendix E.

For comparison, Table 6.1 also presents the parameters of the Mercator projections constructed for the same mapped area with the condition that linear distortions in these projections are minimized according to Airy and Chebyshev's criteria.

A visual representation of changes in extremal linear distortions of two variants of equidistant projections is given in Fig. 6.3.

The graphs in Fig. 6.3 clearly show how distortions are distributed in both projections. The parts of the distortion graphs that are minimal are shown in blue (see formula (6.3)), and the parts of the distortion graphs that are maximum are shown in red.

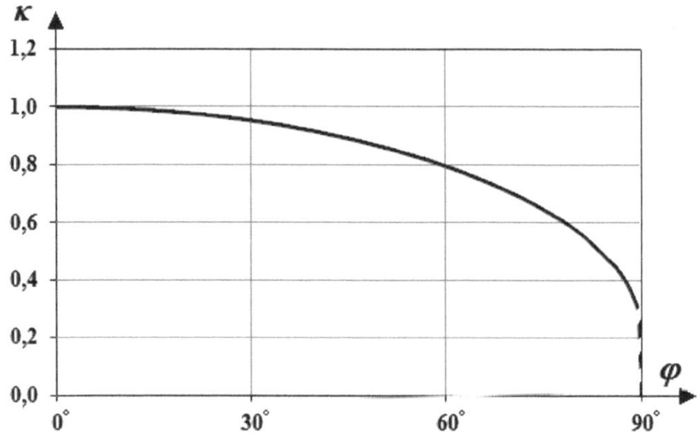

Fig. 6.2 Graph of parameter change κ depending on φ_0 for the WGS84 ellipsoid (with a semi-major axis equal to 1)

Table 6.1 Parameters of the Airy and Chebyshev ideal projections and variants of the Mercator projections constructed for the latitudinal zone with boundary parallels $\varphi_0 = \pm 3°$ for the WGS84 ellipsoid with a semi-major axis equal to 1

Parameter	Projection ideal according to		Mercator projection, best according to	
	Airy	Chebyshev	Airy	Chebyshev
κ	9.9954602E−01	9.986387E−01	9.9954602E−01	9.9863869E−01
$(a+b-1)_{max}$	1.000908569	1.000681118	1.000908569	1.000681118
$(a+b-1)_{min}$	9.9954602E−01	9.9931888E−01	9.9954602E−01	9.9931888E−01
$V_{a\,max}$ (%)	0.091	0.068	0.091	0.068
$V_{b\,min}$ (%)	−0.045	−0.068	−0.045	0.068
$V_{p\,max}$ (%)	0.091	0.068	0.182	0.136
$V_{p\,min}$ (%)	−0.045	−0.068	−0.091	−0.136
Ξ_{Ch} (%)	0.091	0.068	0.091	0.068
Ξ_{A} (%)	0.041	0.046	0.057	0.066

For an ideal Airy projection, regions close to the boundary have less importance than the central region does. As a result, the linear distortion at the central point (−0.045%) is two times less in absolute value than the linear distortion at points on the border (+ 0.091%). Figure 6.3a shows the latitudes of the parallels at which the linear distortion of the Airy ideal projection is equal to + 0.045%. This latitude is ± 2.4°. Similarly, the latitudes of parallels with linear distortions of + 0.068% are equal to ± 2.7°. The area of the territory between parallels with latitudes ± 2.4° is 80%, and that between parallels with latitudes ± 2.7° is 91% of the entire mapping area. Thus, for an Airy ideal projection, linear distortions do not exceed 0.045% for 80% of the mapping area and no more than 0.068% for 91% of the area.

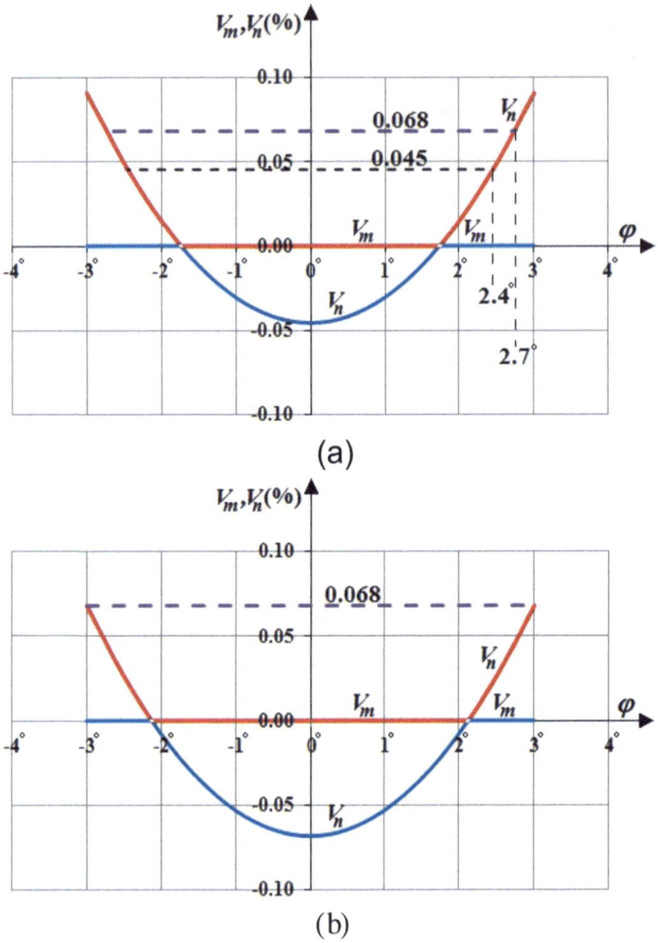

Fig. 6.3 Graphs of changes in extremal linear distortions V_m, V_n of ideal cylindrical projections: **a** according to Airy; **b** according to Chebyshev

For the Chebyshev ideal projection (Fig. 6.3b), all the points of the mapping area are equivalent. Therefore, the minimum linear distortion at the central point (− 0.068%) is equal in absolute value to the maximum linear distortion at the boundary of the region (+ 0.068%).

Since the change in linear distortions in all considered projections is the same, up to values of the second order of smallness, we can write:

$$\min V_b(\text{Ideal according to Airy}) \approx \frac{2}{3}\min V_b(\text{Ideal according to Chebyshev}) \tag{6.16}$$

$$\max V_a(\text{Ideal according to Airy}) \approx \frac{3}{2}\max V_a(\text{Ideal according to Chebyshev}) \tag{6.17}$$

The peculiarity of the mapping area (Fig. 6.1b) leaves a certain imprint on the properties of the Airy ideal projection. Notably, this projection does not fully satisfy the ideal projection theorem, as it was constructed for a doubly connected region. The properties of this projection also do not coincide with those of the Airy ideal projection (see Sect. 6.3). In particular, there is no central point in the projection where extremal distortions are equal. Instead, there are two standard parallels along which there are no linear distortions. Another feature is that one of the extremal scale factors is equal to unity not only at the boundary of the mapping area but throughout its entire territory.

6.2 Gauss–Kruger Projection and Its Properties

The Gauss–Kruger projection is one of the most successful projections known to mankind (Annoni et al. 2003; Hager et al. 1986). Created in the nineteenth century by K.F. Gauss for processing the Hanoverian triangulation and later improved by Louis Kruger in the early twentieth century, it serves as the basis for large-scale mapping in most countries worldwide.

The Gauss–Kruger projection has several properties that make it ideal for large-scale mapping. These properties are listed in Table 6.2 and are divided into two parts. The first part presents properties common to all conformal projections. The second part contains only those properties that are not related to conformity and are unique properties specific to the Gauss–Kruger projection itself.

A completely reasonable question arises: why did K.F. Gauss choose this particular projection to solve geodetic problems? What property of the Gauss–Kruger projection turned out to be decisive for this choice? At first glance, such a key property is the nondistortion of angles, since most of all measurements in triangulation are measurements of angles. In conformal projections, angles are not distorted, so projecting the measured angles of a triangulation network onto a plane does not require corrections for angle distortion. Apparently, this consideration, along with the profound admiration for the greatness of K.F. Gauss, compelled surveyors to use the Gauss-Kruger projection or other conformal projections for geodetic purposes. Examples include the Lambert projection (conic conformal) and the Hipparchus stereographic projection (azimuthal conformal) (Snyder 1987, pp. 104–106, 154–157, 203–207; Lapaine and Tutic 2008).

However, although the angles in the conformal projections are not distorted, corrections are still introduced before processing:

- Correction for the transition from normal sections to geodesic lines on the surface of the ellipsoid;

Table 6.2 Properties of the Gauss–Kruger projection

1. Properties belonging to all conformal projections	2. Individual properties of the projection
1.1 The angles are not distorted	2.1 The Earth's ellipsoid is divided into zones by meridians. They can be zones of the same size (6-degree or 3-degree), completely covering the entire Earth's ellipsoid, or they may be one or two zones, the sizes of which are selected individually for the territory of the country
1.2 The similarity of infinitesimal figures is preserved	2.2 Each zone is projected onto the plane separately
1.3 The linear scale factor does not depend on direction	2.3 The central meridian of the zone is depicted by a straight line coinciding with the N axis
1.4 Area distortions are twice as large as linear ones	2.4 The linear scale factor along the central meridian is a constant value
1.5 The mapping functions (N, E) are conjugate harmonic functions	2.5 The equator is depicted as a straight line coinciding with the E axis
1.6 The functions $ra\cos\gamma$ and $-ra\sin\gamma$ are conjugate harmonic functions	2.6 Projection formulae do not depend on the absolute longitude of the point on the ellipsoid. They are unified and depend on the difference between the longitude of the projection point and the longitude of the central meridian
1.7 The functions $\ln(ra)$ and $(-\gamma)$ are conjugate harmonic functions	2.7 All functional parameters of the projection are defined as the sum of the first terms of an infinite series

- Correction for the transition from azimuths to grid azimuths of geodesic lines on the plane in the projection, called planar grid convergence of meridians;
- Correction for the curvature of the image of a geodesic line on a plane in projection for the transition from the grid azimuth of the geodesic line to the grid azimuth of the chord connecting the starting and ending points of the geodesic line on the plane. This correction is called arc-to-chord correction.

  If the projection is not conformal, then one more correction is added to this list of corrections:
- Correction for distortion of the azimuth of the direction when transitioning from the surface to the plane.

Therefore, the property of conformity is not key. It does not provide ideal benefits when working with measured angles.

Let us examine in more detail such a property of projection as the use of conjugate harmonic functions to determine mapping functions (property 1.5, Table 6.2).

Suppose that polynomials composed of infinite series are used to determine the projection. The two mapping functions of nonconformal projections require two polynomials for their definition:

$$N = \sum_{k=0}^{J} \sum_{j=0}^{J} c_{kj} q^k \lambda^j, E = \sum_{k=0}^{J} \sum_{j=0}^{J} d_{kj} q^k \lambda^j, \tag{6.18}$$

where c_{kj}, d_{kj} are constant projection parameters.

For a conformal projection, considering that the mapping functions are conjugate harmonic functions, the sums of the first terms of the series of these functions can be written as follows:

$$N = \sum_{j=0}^{J} \left(c'_j P_j + d'_j Q_j \right), E = \sum_{j=0}^{J} \left(c'_j Q_j - d'_j P_j \right), \tag{6.19}$$

where c'_j, d'_j are constant projection parameters and where P_j, Q_j are harmonic polynomials determined by the following formulae (Bugayevskiy and Snyder 1995, p. 166):

$$P_0 = 1, Q_0 = 0, P_j = P_{j-1} \cdot q - Q_{j-1} \cdot \lambda, Q_j = P_{j-1} \cdot \lambda + Q_{j-1} \cdot q, \tag{6.20}$$

Obviously, according to formulae (6.19, 6.20), the number of coefficients necessary to determine the nonconformal projection is an order of magnitude greater than the number of coefficients necessary to determine the conformal projection. This is shown more clearly in Fig. 6.4.

In Fig. 6.4, each coefficient corresponds to a square. The number of coefficients depends on the degree J of the polynomial. For example, a polynomial of degree $J = 1$ for the mapping function N of a nonconformal projection contains coefficients c_{01}, c_{10}; a similar polynomial for the mapping function E also contains two coefficients d_{01}, d_{10}. The total number of coefficients for polynomials of degree $J = 1$ used in mapping functions of nonconformal projections is 4. For the conformal projection, polynomials of degree $J = 1$ for the functions N, E have only two coefficients c'_1, d'_1.

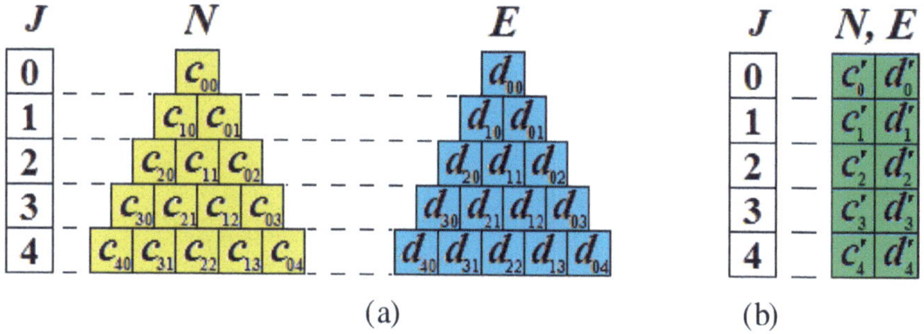

Fig. 6.4 Visual representation of the dynamics of increasing the number of coefficients of mapping functions, presented in the form of polynomials, with increasing degree: **a** for nonconformal projections; **b** for conformal projections

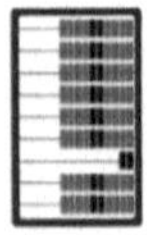

According to Fig. 6.4 and formulae (6.18, 6.19), determining the mapping functions of nonconformal projections in the form of polynomials of degree J requires finding $(J+1)(J+2)$ coefficients. In contrast, defining mapping functions as polynomials for conformal projections only requires $2(J+1)$ coefficients. Currently, given the availability of computers, the differences between formulae (6.18) and (6.19) are not significant. However, in the nineteenth century, when computers looked like the ones shown at the beginning of this paragraph and all calculations were carried out almost manually, the advantages of formulae (6.19) were undeniable. Moreover, the linear distortions of both conformal and nonconformal projections, as indicated in Sect. 6.1 (see Table 1.1), are almost identical. Therefore, it is reasonable to believe that the minimum number of coefficients for mapping functions was decisive in choosing conformal projections.

To determine the parameters of his projection, K.F. Gauss chose not harmonic polynomials but rather a simpler and at the same time ingenious solution, according to which all conjugate harmonic functions of the projection (see Table 6.2, properties 1.5–1.7) are determined using the Maclaurin formula, which is applied to the corresponding analytical function. Namely:

$$F(q+il) = F(q) + \frac{1}{1!}\frac{dF}{dq}(q)(il) + \frac{1}{2!}\frac{d^2F}{dq^2}(q)(il)^2 + \frac{1}{3!}\frac{d^3F}{dq^3}(q)(il)^3 + \cdots, \tag{6.21}$$

where i is the imaginary unit and where l is the difference between the longitude of the point and the longitude of the central meridian of the zone:

$$l = \lambda - \lambda_0. \tag{6.22}$$

When the mapping functions are defined, the analytical function is as follows:

$$F(q+il) = N_{G-K} + iE_{G-K}, \tag{6.23}$$

or,

$$N_{G-K} + iE_{G-K} = F(q) + \frac{1}{1!}\frac{dF}{dq}(q)il - \frac{1}{2!}\frac{d^2F}{dq^2}(q)l^2 - \frac{1}{3!}\frac{d^3F}{dq^3}(q)il^3 + \cdots, \tag{6.24}$$

Separating the real and imaginary parts, we obtain:

$$\begin{aligned} N_{G-K} &= F(q) - \tfrac{1}{2!}\tfrac{d^2F}{dq^2}(q)l^2 + \cdots \\ E_{G-K} &= \tfrac{1}{1!}\tfrac{dF}{dq}(q)l - \tfrac{1}{3!}\tfrac{d^3F}{dq^3}(q)l^3 + \cdots, \end{aligned} \tag{6.25}$$

To determine all conjugate harmonic functions of the Gauss–Kruger projection (including the functions in equalities (6.25)), property 2.4 from Table 6.2 is crucial. According to this property, the linear scale factor on the central meridian is equal to a constant m_0. This means that the coordinate N for the points of the central meridian is equal to the length of the central meridian itself from a selected fixed point, such as a point on the equator, multiplied by the constant m_0, i.e.,

$$N_{G-K}|_{\lambda=0} = F(q) = m_0 \int_0^q r dq = m_0 \cdot S. \tag{6.26}$$

The derivatives of formulae (6.25) are determined via Eq. (6.26) as follows:

$$\begin{array}{l} \frac{dF}{dq}(q) = m_0 \cdot r \\ \frac{d^2F}{dq^2}(q) = m_0 \cdot r_q \\ \ldots \\ \frac{d^jF}{dq^j} = m_0 \cdot r_{q(j-1)} \end{array}. \tag{6.27}$$

In (6.27), $r_q, \ldots r_{q(j-1)}$ are the first, second, and so on up to the $(j-1)$-th derivatives of the function r with respect to isometric latitude q. The formulae for calculating these derivatives are presented in Appendix A.

Taking into account (6.26, 6.27), formulae (6.25) can be written as follows:

$$\begin{array}{l} N_{G-K} = m_0\left(S - \frac{1}{2!}r_q l^2 + \cdots\right) \\ E_{G-K} = m_0\left(\frac{1}{1!}rl - \frac{1}{3!}r_{qq} l^3 + \cdots\right) \end{array}. \tag{6.28}$$

The substitution of these derivatives after identity transformations gives the following well-known equalities (Deakin and Hunter 2010; Deakin et al. 2012; Li et al. 2016; Masaharu 2017):

$$\begin{array}{l} N_{G-K} = m_0\left[S + \frac{l^2}{2} r \sin\varphi \left\{1 + \frac{l^2 \cos^2\varphi}{12}\left(5 - t^2 + 9\eta^2 + 4\eta^4\right),\right.\right. \\ \left.\left. + \frac{l^4 \cos^4\varphi}{360}\left(61 - 58t^2 + t^4\right)\right\}\right] \end{array} \tag{6.29}$$

$$\begin{array}{l} E_{G-K} = m_0 rl\left[1 + \frac{l^2 \cos^2\varphi}{6}\left(1 - t^2 + \eta^2\right),\right. \\ \left. + \frac{l^4 \cos^4\varphi}{120}\left(5 - 18t^2 + t^4 + 14\eta^2 - 58\eta^2 t^2\right)\right] \end{array} \tag{6.30}$$

Where

$$\eta^2 = e'^2 \cos^2\varphi, t = \tan\varphi. \tag{6.31}$$

For arbitrary mapping functions N, E, we can also write formulae similar to those in (6.25), namely:

$$\begin{aligned} N &= F_0(q) - \tfrac{1}{2!}F_2(q)l^2 + \tfrac{1}{4!}F_4(q)l^4 \dots \\ E &= \tfrac{1}{1!}F_1(q)l - \tfrac{1}{3!}F_3(q)l^3 + \cdots \end{aligned}. \tag{6.32}$$

However, in this case, the functions $F_0(q), \dots F_3(q) \dots$ are no longer derivatives of a single function. In general, these functions are independent of each other.

Thus, the Gauss–Kruger projection possesses a set of properties, each of which is not unique, but together, they have proven so successful that since the nineteenth century, it has been widely used for large-scale mapping in almost all countries of the world. However, one important drawback cannot be ignored.

The widespread use of the Gauss–Kruger projection for cadastral purposes has revealed that the crucial property of the projection is not the distortions of angles or line lengths but the distortions of areas. The area of the land parcel determines its value, land tax, rental amount, and so on. Moreover, conformal projections have large area distortions, which are twice as large as linear distortions (see Sect. 1.10). Recent research (Novikova et al. 2023) demonstrates that neglecting area distortions of land parcels has led to issues in their accounting in Ukraine.

There are several ways to combat area distortions when calculating the official area of a parcel, i.e., the area specified in its documentation:

1. The official area of the parcel is calculated not by rectangular coordinates on a plane in projection but by geodetic coordinates on the surface of the Earth's ellipsoid.
2. The official area of the parcel is calculated via rectangular coordinates on a plane, but corrections for area distortions are subsequently introduced (LEA 2021).
3. Conformal projection is replaced with a nonconformal projection, in which area distortions are either zero or minimized.

The third method is the most radical and, as a rule, requires a transition to a new national coordinate system, which involves significant financial costs. Usually, this method is not even considered by specialists, since, according to (Timar et al. 2011), cadastral work is expensive, and the state is unlikely to fund its repetition in a new coordinate system.

Moreover, using the first method to solve the problem of area distortion turns cadastral plans into simple visual aids, making it impossible to accurately calculate land parcel areas using them.

The second method of solving the problem allows us to preserve the significance of the cadastral plan, but it should always be remembered that the areas defined by rectangular coordinates have an error and must be corrected.

As shown below, using the theory of best projections allows for a relatively painless transition from the Gauss–Kruger projection to a similar projection with the same linear distortions but without area distortions.

6.3 Gauss–Kruger Projection with Minimal Linear Distortions

The Gauss–Kruger projection is not the best conformal projection for the longitudinal zone since it does not satisfy Chebyshev's theorem for the best conformal projections: the linear scale factor on the boundary meridians is not equal to a constant value.

The following question arises: how close is the Gauss–Kruger projection to the best projection? To answer this question, we write Eq. (6.21) for the conjugate harmonic functions $\ln(rm)$ and $(-\gamma)$:

$$\ln(rm) - i\gamma = F''(q) + \frac{1}{1!}\frac{dF''}{dq}(q)(il) + \frac{1}{2!}\frac{d^2F''}{dq^2}(q)(il)^2 + \frac{1}{3!}\frac{d^3F''}{dq^3}(q)(il)^3 + \cdots, \tag{6.33}$$

or, separating the real and imaginary parts:

$$\begin{aligned} \ln(rm) &= F''(q) - \tfrac{1}{2!}\tfrac{d^2F''}{dq^2}(q)l^2 + \cdots \\ \gamma &= -\tfrac{1}{1!}\tfrac{dF''}{dq}(q)l + \tfrac{1}{3!}\tfrac{d^3F''}{dq^3}(q)l^3 + \cdots \end{aligned} . \tag{6.34}$$

As in the previous case, the function F'' and its derivatives for the Gauss–Kruger projection are determined from the condition that, on the central meridian, the linear scale factor, and therefore, the logarithm of the linear scale factor, is equal to a constant.

Since the Gauss–Kruger projection is used with a wide variety of linear scale factors along the central meridian, among all the variants of this projection, we choose one for which $\ln m$ has a minimum on the central meridian equal to $-k_{G-K}$ and a maximum at the point of intersection of the equator with the boundary meridian, equal to $+k_{G-K}$. In this case, the maximum linear distortion, determined by the second formula of Eq. (2.24), is equal to the absolute value of the minimum distortion.

Thus, the function F'' and its derivatives are equal:

$$F''(q) = \ln(rm_0) = \ln r + \ln m_0 = \ln r - k_{G-K}, \tag{6.35}$$

$$\begin{aligned} \frac{dF''}{dq} &= \tau_q \\ \frac{d^2F''}{dq^2} &= \tau_{qq} \\ &\cdots \\ \frac{d^jF''}{dq^j} &= \tau_{jq} \end{aligned} , \tag{6.36}$$

where τ is the logarithm of the coefficient of the metric shape of the surface. The working formulae for determining the derivatives of a function τ are presented in Appendix A.

Using equalities (6.35) and (6.36), we obtain:

$$\ln m = \ln r - k_{G-K} - \frac{1}{2!}\frac{d^2F''}{dq^2}l^2 + \frac{1}{4!}\frac{d^4F''}{dq^4}l^4 - \cdots \ln r, \quad (6.37)$$

or,

$$\ln m = -k_{G-K} - \frac{1}{2!}\frac{d^2F''}{dq^2}l^2 + \frac{1}{4!}\frac{d^4F''}{dq^4}l^4 - \cdots . \quad (6.38)$$

For the Gauss–Kruger projection, the linear scale factor and the logarithm of the linear scale factor reach a maximum at the point of intersection of the boundary meridian and the equator, i.e., at a point with coordinates $q = 0, l = \pm l_0$. At these points, the derivatives of the function τ, according to formulae (A.12, A.14, A.16) of Appendix A, are equal to:

$$\begin{aligned} \tau_{qq}|_{q=0} &= -(1 + e'^2) \\ \tau_{4q}|_{q=0} &= 2(1 + e'^2)^2(1 + 2e'^2) \\ \tau_{6q}|_{q=0} &= -8(1 + e'^2)^3(2 + 11e'^2 + 11e'^4) \end{aligned}, \quad (6.39)$$

and the logarithm of the linear scale factor is equal to k_{G-K}, namely:

$$\ln m_{\max} = -k_{G-K} + \frac{1}{2!}(1 + e'^2)l_0^2 + \frac{2}{4!}(1 + e'^2)^2(1 + 2e'^2)l_0^4 + \cdots = k_{G-K}. \quad (6.40)$$

Let us find k_{G-K} from this equality.

$$k_{G-K} = \frac{1}{2 \cdot 2!}(1 + e'^2)l_0^2 + \frac{2}{2 \cdot 4!}(1 + e'^2)^2(1 + 2e'^2)l_0^4 + \cdots . \quad (6.41)$$

For a six-degree zone, the value of l_0 is:

$$l_0 = \pm\frac{3^\circ \cdot \pi}{180^\circ}. \quad (6.42)$$

The numerical value of k_{G-K} for the six-degree zone is presented in Table 6.3.

Table 6.3 Parameters of Gauss–Kruger projection and the best conformal one for the six-degree zone of the WGS84 ellipsoid

Parameter	Numerical value
k_{G-K}	0.0006903
k_{best}	0.0006884
$k_{G-K} - k_{\text{best}}$	0.0000019

6.4 The Best Conformal Projection for a Narrow Longitudinal Zone

The logarithm of the linear scale factor for the best conformal projection of a narrow longitudinal zone can be described using the first formula of the system (6.34). However, in this case, the function F'' is not determined by the value of the function $\ln(rm)$ on the central meridian but by its value on the boundary meridian. The function F'' is sought in the form of a polynomial in powers of isometric latitude q:

$$F'' = \theta_0 + \theta_2 q^2 + \theta_4 q^4 + \cdots, \tag{6.43}$$

where $\theta_0, \theta_2, \theta_4$ are some constants.

Let us find the derivatives of the function F'':

$$\begin{aligned} \frac{d^2F''}{dq^2} &= 2!\theta_2 + 4\cdot 3\theta_4 q^2 + \cdots \\ \frac{d^4F''}{dq^4} &= 4!\theta_4 + \cdots \end{aligned} . \tag{6.44}$$

Substituting the function F'' and its derivatives into the first formula of the system (6.34) leads to:

$$\ln m = \theta_0 + \theta_2 q^2 + \theta_4 q^4 + \cdots - \frac{1}{2!}\left(2!\theta_2 + 4\cdot 3\theta_4 q^2 + \cdots\right)l^2 + \frac{1}{4!}(4!\theta_4 + \cdots)l^4 - \tau. \tag{6.45}$$

Let us write the Maclaurin series for the function τ:

$$\tau + \tau_0 + \frac{1}{2!}(\tau_{qq})_0 q^2 + \frac{1}{4!}(\tau_{4q})_0 q^4 + \cdots . \tag{6.46}$$

For an ellipsoid similar to the Earth's ellipsoid but with a semi-major axis equal to one, the value of τ_0 is 0. The numerical values of the derivatives of the function τ at $q = 0$ are presented in equalities (6.39). Hence,

$$\tau = -\frac{1}{2!}(1 + e'^2)q^2 + \frac{1}{4!}2(1 + e'^2)^2(1 + 2e'^2)q^4 + \cdots . \tag{6.47}$$

Consequently, taking into account (6.45, 6.47), we get.

$$\begin{aligned} \ln m = {} & \theta_0 + \theta_2 q^2 + \theta_4 q^4 + \cdots - \frac{1}{2!}(2!\theta_2 + 4\cdot 3\theta_4 q^2 + \cdots)l^2 + \frac{1}{4!}(4!\theta_4 + \cdots)l^4 \\ & + \frac{1}{2!}(1 + e'^2)q^2 - \frac{1}{4!}2(1 + e'^2)^2(1 + 2e'^2)q^4 + \cdots . \end{aligned} \tag{6.48}$$

For the best projection, the logarithm of the maximum linear scale factor reaches a maximum equal to k_{best} at the boundary meridians and remains constant.

Thus, formula (6.49) for the boundary meridians is rewritten as follows:

$$k_{\text{best}} = \theta_0 + \theta_2 q^2 + \theta_4 q^4 + \cdots - \frac{1}{2!}(2!\theta_2 + 4 \cdot 3\theta_4 q^2 + \cdots) l_0^2 + \frac{1}{4!}(4!\theta_4 + \cdots) l_0^4 + \frac{1}{2!}(1 + e'^2) q^2 - \frac{1}{4!} 2(1 + e'^2)^2 (1 + 2e'^2) q^4 + \cdots . \tag{6.49}$$

Equality (6.49) is a function of isometric latitude, which can be expressed as follows:

$$B_0 + B_2 q^2 + B_4 q^4 + \cdots = 0, \tag{6.50}$$

Where

$$\begin{aligned} B_0 &= \theta_0 - \theta_2 l_0^2 + \theta_4 l_0^4 - \cdots - k_{best} \\ B_2 &= \theta_2 - \tfrac{1}{2!} 4 \cdot 3\theta_4 l_0^2 + \cdots + \tfrac{1}{2!}(1 + e'^2) . \\ B_4 &= \theta_4 + \cdots - \tfrac{1}{4!} 2(1 + e'^2)^2 (1 + 2e'^2) \end{aligned} \tag{6.51}$$

For the function ln$m$ to equal the value k_{best} at all points on the boundary of the region, the coefficients $B_j (j = 0, 2, \dots)$ must be equal to zero. Setting these coefficients to zero yields a system of linear equations. Limiting the degree of the polynomial (6.50) to 4 results in a linear system of three equations with four unknowns: $B_0, B_2, B_4, k_{\text{best}}$. The fourth equation is obtained by requiring that the function ln$m$ reaches its minimum, equal to $(-k_{\text{best}})$, at the central point of the projection with coordinates $q = 0, l = 0$. Substituting these coordinates and the value of the minimum of the function ln$m$ into formula (6.48) yields:

$$\ln m_{\min} = -k_{\text{best}} = \theta_0 . \tag{6.52}$$

Combining Eq. (6.52) with the system of Eqs. (6.51) gives the following system of three equations with three unknowns:

$$\begin{aligned} &k_{\text{best}} = -\tfrac{1}{2}\theta_2 l_0^2 + \tfrac{1}{2}\theta_4 l_0^4 \\ &\theta_2 - \tfrac{1}{2!} 4 \cdot 3\theta_4 l_0^2 + \tfrac{1}{2!}(1 + e'^2) = 0 , \\ &\theta_4 - \tfrac{1}{4!} 2(1 + e'^2)^2 (1 + 2e'^2) = 0 \end{aligned} \tag{6.53}$$

The substitution of the values of θ_2, θ_4 from the second and third equations into the first formula of the system (6.53) gives the following formula for determining k_{best}:

$$k_{\text{best}} = \frac{1}{2 \cdot 2!} l_0^2 \left(1 + e'^2\right) + \frac{1}{4!} l_0^4 (1 + e'^2)^2 (1 + 2e'^2) - \frac{1}{4} l_0^4 (1 + e'^2)^2 \left(1 + 2e'^2\right). \tag{6.54}$$

The comparison of formulae (6.41) and (6.54) allows us to conclude that the value of k_{best} is less than k_{G-K}. The difference between these values is as follows:

$$k_{G-K} - k_{\text{best}} = \frac{1}{4} l_0^4 (1 + e'^2)^2 \left(1 + 2e'^2\right) \tag{6.55}$$

If the quantity l_0^2 is of the first order of smallness, then the quantity l_0^4 is of the second order of smallness.

In this case, a comparison of equalities (6.41) and (6.54) allows us to conclude that the linear distortions of the Gauss–Kruger projection and the best projection for the longitudinal zone have one order of smallness. Differences occur only in quantities of the second order of smallness.

Table 6.3 shows the characteristics of both projections for a six-degree zone on the WGS84 ellipsoid.

On the basis of the data in Table 6.3, we can conclude that the Gauss–Kruger projection has linear distortions almost equal to those of the best projection.

References

Abee MD (2019) The Mercator Projection: its Uses, Misuses, and its Association with Scientific Information and the Rise of Scientific Societies. Dissertation, University of North Carolina. p 309. https://libres.uncg.edu/ir/uncg/f/Abee_uncg_0154D_12893.pdf

Alegria *MF, Daveau S, Garcia JC, Relaño F (2007) Portuguese Cartography in the Renaissance. In Woodward, David (ed.).* Cartography in the European Renaissance. *The History of Cartography. Vol. 3. University of Chicago Press, p 1002.* ISBN 978-0-226-90733-8

Annoni A, Luzet C, Gubler E, Ihde J (2003) Map Projections for Europe. Institute for Environment and Sustainability. European Communities, p 131. http://mapref.org/LinkedDocuments/MapProjectionsForEurope-EUR-20120.pdf

Ben Hadj Salem A, Commiot J (2021) Elements de Cartographie Mathematique (v3). Ingénieurs Généraux Géographes. https://www.researchgate.net/publication/354511206_Elements_of_Mathematical_Cartography_v3

Buchroithner MF, Pfahlbusch R (2016) Geodetic grids in authoritative maps – new findings about the origin of the UTM Grid. Cartography and Geographic Information Science, January 2016: 1–15 https://doi.org/10.1080/15230406.2015.1128851. https://doi.org/10.1080/15230406.2015.1128851

Bugayevskiy LM, Snyder JP (1995) Map projections. Reference Manual. Taylor & Francis Ltd., London, p 328. https://doi.org/10.1201/b16431

Daners D (2012) The Mercator and stereographic projections, and many in between. The American Mathematical Monthly 119(3): 199–210. https://www.maths.usyd.edu.au/u/daners/publ/abstracts/mercator/mercator.pdf

De Jong SH (1968) The Skew Mercator Projection in Rectangular Plane Coordinate Systems, Dissertation, The Ohio State University, p 160. https://etd.ohiolink.edu/acprod/odb_etd/ws/send_file/send?accession=osu1486647101628433&disposition=inline

Deakin RE, Hunter MN, Karney CFF (2010) The Gauss–Kruger Projection. Paper presented at Victorian Regional Survey Conference, Warrnambool, 10–12 September 2010, p 20. https://www.mygeodesy.id.au/documents/Gauss-Krueger%20Warrnambool%20Conference%20V2.pdf

Deakin RE, Hunter MN, Karney CFF (2012) A Fresh Look at the UTM Projection: Karney-Krueger equations. Paper presented at Surveying and Spatial Sciences Institute (SSSI) Land Surveying Commission National Conference, Melbourne, 18–21 April 2012, p 19. https://www.mygeodesy.id.au/documents/A%20fresh%20look%20at%20the%20UTM%20projection%20-%20the%20Karney-Krueger%20equations%20V2.pdf

Fabregat M (2017) Conformal Cartographic Representations. Facultat de Matematiques i Informatica Universitat de Barcelona. Barcelona, p 103. https://diposit.ub.edu/dspace/bitstream/2445/121898/2/memoria.pdf. Accessed 29 June 2017

Grafarend EW, Krumm FW (2006) Map Projections. Cartographic Information Systems. Springer-Verlag Berlin Heidelberg, p 713. https://books.google.com.ua/books/about/Map_Projections.html?id=Rg3Lc8omIXMC&redir_esc=y

Hager JW, Fir LL, Jacks SS, Hiji DR (1986) Datums, Ellipsoids, Grids, and Grid Reference Systems. DMA Technical Manual: TM 8358.1. The Defense Mapping Agency. June 1986, p 173. https://apps.dtic.mil/sti/pdfs/ADA247651.pdf

Korn GA, Korn TM (1968) Mathematical Handbook: for Scientists and Engineers. McGraw-Hill Book Company. New York, San Francisco, Toronto, London, Sydney. p 832. https://archive.org/details/mathematicalhand0000gran/page/n7/mode/2up

LEA (2021) Anweisung zur Erhebung von Daten für das Liegenschaftskataster - Liegenschaftsdatenerhebungsanweisung. Stand: 12. Januar 2021. Hessisches Ministerium für Wirtschaft, Energie, Verkehr und Landesentwicklung. https://hvbg.hessen.de/sites/hvbg.hessen.de/files/2022-08/liegenschaftsdatenerhebungsanweisung.pdf

Li Z, Bian S, Liu Q, Li H, Chen C, Hu Y (2016) Nonzonal Expressions of Gauss- Kruger Projection in Polar Regions. ISPRS Annals of the Photogrammetry Remote Sensing and Spatial Information Sciences III-4: 11–15. https://doi.org/10.5194/isprs-annals-III-4-11-2016. https://isprs-annals.copernicus.org/articles/III-4/11/2016/isprs-annals-III-4-11-2016.pdf

Masaharu H (2017) Development in understanding of Gauss–Krüger projection and its outcomes. Proceedings of the International Cartographic Association, 1:1–4. https://www.researchgate.net/publication/325187827_Development_in_understanding_of_Gauss-Kruger_projection_and_its_outcomes

Novikova E, Palamar A, Lopunov R (2023) The dual choice of geodetic horizontal reference systems for Ukraine. Geod Geodyn 14(3):254–264. https://doi.org/10.1016/j.geog.2022.11.001

Osborn P (2013) The Merkator Projections. The Normal and Transverse Mercator Projections on the Sphere and the Ellipsoid with Full Derivations of all Formulae. Edinburgh, p 254. https://doi.org/10.5281/zenodo.35392. https://zenodo.org/records/35561

Snyder JP (1987) Map projectionsa— working manual, U.S. Geological Survey professional paper 1395, USPOGO, Washington D.C, p 397. https://doi.org/10.3133/pp1395

Timar G, Baiocchi V, Lelo K (2011) Geodetic Datums of the Italian Cadastral Systems. Geographia Technica, 2011 1:82–90. http://technicalgeography.org/pdf/1_2011/08_gabor_timar_valerio_baio.pdf

Tutic D, Lapaine M (2008) Stereographic map projection of Croatia. Paper presented at the 13th international conference on geometry and graphics (ICGG 2008): Proceedings/Weiss, Gunter (ur.). Dresden: ISGG and Technische Universitat Dresden. https://www.researchgate.net/publication/228653585_Stereographic_Map_Projection_of_Croatia

Chapter 7
Arithmetic Mean Principle for the Gauss–Kruger Projection

In fact, the Gauss–Kruger projection is the best conformal projection, not for the area limited by two meridians but for the area limited by two curves on the ellipsoid, which serve as parallels when transitioning from a direct to a transverse projection and are determined by the following equation:

$$\ln m = -k_{G-K} - \frac{1}{2!}\tau_{qq}l_0^2 + \frac{1}{4!}\tau_{qqqq}l_0^4 + \cdots = k_{G-K} = \text{const.} \tag{7.1}$$

These curves on the ellipsoid are schematically shown by the white dashed lines in Fig. 7.1. The meridional zone for which the projection is being constructed is shown in green, with its central meridian in red. As shown in Fig. 7.1, the region bounded by the white dashed lines, for which the Gauss–Kruger projection is the best conformal projection, does not coincide with the region bounded by two meridians. The fundamental difference between these regions is clearly visible. The first region is doubly connected, whereas the second region is simply connected. The regions touch only by two boundary points, which are the points where the boundary meridians intersect with the equator.

As shown in Sect. 7.6.1, the direct cylindrical projection, which is the best according to the Airy criterion, is the best for a doubly connected mapping area. Therefore, this projection does not satisfy the theorem on Airy ideal projections. However, as shown below, the arithmetic mean principle (see Sect. 4.5) can be **applied to** doubly connected regions. According to **this principle,** if at least two of the **following** three projections are known:

- the best conformal projection;
- the Airy ideal projection;
- the best projection from the set of close-to-equal-area projections.

then the third projection **can be easily** determined **using** the arithmetic mean principle.

E. Novikova, *Best Map Projections*, https://doi.org/10.1007/978-3-031-78334-0_7

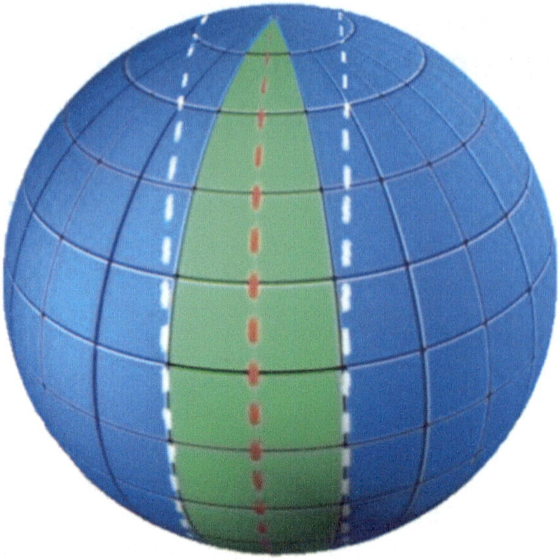

Fig. 7.1 Mapping area on the ellipsoid (shown as white dashed lines) for which the Gauss–Kruger projection is the best conformal projection

Let us examine the application of this principle to the mapping area under consideration.

7.1 Airy Ideal Projection Associated with Gauss–Kruger Projection

Let us rewrite the Formulae (2.7) as follows:

$$N_\lambda - E_q = r(a+b)\sin\beta = r(a+b-1)\sin\beta + r\sin\beta, \tag{7.2}$$

$$N_q + E_\lambda = r(a+b)\cos\beta = r(a+b-1)\cos\beta + r\cos\beta. \tag{7.3}$$

We search for mapping functions of the Airy ideal projection in the form of polynomials in powers of l, namely:

$$\begin{aligned} N' &= \sum_{j=0}^{J} N'_{2j} l^{2j} \\ E' &= \sum_{j=0}^{J} E'_{2j+1} l^{2j+1}, \end{aligned} \tag{7.4}$$

where N'_{2j}, E'_{2j+1}, $(j = 0, 1, \ldots J)$ are some functions that depend only on latitude.

Like the original Gauss–Kruger projection functions, N_{G-K}, E_{G-K}, the function N' is even, and the function E' is odd with respect to the variable l. This property of the functions is dictated by the symmetry of the mapping area along the N, E axes.

Let us find the derivatives of the functions N', E':

$$\begin{aligned} N'_q &= \sum_{j=0}^{J} N'_{2jq} l^{2j} \\ N'_\lambda &= \sum_{j=0}^{J} 2j \cdot N'_{2j} l^{2j-1}, \end{aligned} \tag{7.5}$$

$$\begin{aligned} E'_q &= \sum_{j=0}^{J} E'_{(2j+1)q} l^{2j+1} \\ E'_\lambda &= \sum_{j=0}^{J} (2j+1) E'_{2j+1} l^{2j}. \end{aligned} \tag{7.6}$$

Substituting these derivatives into (7.2–7.3) yields:

$$N'_\lambda - E'_q = \sum_{j=1}^{J} \left(2jN'_{2j} - E'_{2j-1}\right) l^{2j-1}, \tag{7.7}$$

$$N'_q + E' = \sum_{j=0}^{J} \left(N'_{2j} + (2j+1)E'_{2j+1)}\right) l^{2j+1}. \tag{7.8}$$

As indicated in Sect. 4.5, the relationships between the three abovementioned projections are written as follows:

$$\ln m|_{\text{best conformal}} = \ln(a+b-1)|_{\text{ideal to Airy}} = \ln(1+L)|_{\text{best close-to-equal-area}}, \tag{7.9}$$

$$\gamma = \beta|_{\text{best conformal}} = \beta|_{\text{ideal to Airy}} = \beta|_{\text{best close-to-equal-area}}. \tag{7.10}$$

To determine the conjugate harmonic functions $rm\cos\gamma$ and $-rm\sin\gamma$ of the Gauss–Kruger projection, we use Formula (6.21), namely:

$$rm\cos\gamma - irm\sin\gamma = F'''(q) + \frac{1}{1!}\frac{\mathrm{d}F'''}{\mathrm{d}q}(q) \cdot il + \frac{1}{2!}\frac{\mathrm{d}^2F'''}{\mathrm{d}q^2}(q) \cdot (il)^2 + \cdots, \tag{7.11}$$

where the real and imaginary parts are separated as follows:

$$\begin{aligned} rm\cos\gamma &= F'''(q) - \tfrac{1}{2!}\tfrac{\mathrm{d}^2F'''}{\mathrm{d}q^2}(q) \cdot l^2 + \cdots \\ rm\sin\gamma &= -\tfrac{1}{1!}\tfrac{\mathrm{d}F'''}{\mathrm{d}q}(q) \cdot l + \tfrac{1}{3!}\tfrac{\mathrm{d}^3F'''}{\mathrm{d}q^3}(q) \cdot l^3 - \cdots. \end{aligned} \tag{7.12}$$

As with the definition of other pairs of conjugate harmonic functions in the Gauss–Kruger projection, the function F''' and its derivatives are determined by the condition that the linear scale factor on the central meridian is equal to a constant. Taking into account Formula (6.38), we write:

$$m_{\min} = m|_{l=0} = m_0 = \exp(-k_{G-K}) \approx 1 - k_{G-K}. \tag{7.13}$$

The numerical value of k_{G-K} is presented in Table 6.3.

Therefore, the function F''' and its derivatives are equal to:

$$F'''(q) = rm_0, \quad \frac{dF'''}{dq}(q) = m_0 r_q, \quad \frac{d^2F'''}{dq^2}(q) = m_0 r_{qq}, \quad \frac{d^jF'''}{dq^j}(q) = m_0 r_{qj}. \tag{7.14}$$

Substituting (7.14) into (7.12) yields:

$$\begin{aligned} rm\cos\gamma &= m_0\left(r - \tfrac{1}{2!}r_{qq}l^2 + \tfrac{1}{4!}r_{4q}l^4 + \cdots\right) \\ rm\sin\gamma &= m_0\left(-r_q l + \tfrac{1}{3!}r_{3q}l^3 - \tfrac{1}{5!}r_{5q}l^5 + \cdots\right). \end{aligned} \tag{7.15}$$

To determine $\sin\gamma, \cos\gamma$, one can use the well-known expansions of these functions in the Maclaurin series (Dwight 1961, p. 79):

$$\begin{aligned} \sin\gamma &= \sum_0^\infty \tfrac{1}{(2j+1)!}(-1)^j\gamma^{(2j+1)} \\ \cos\gamma &= \sum_0^\infty \tfrac{1}{(2j)!}(-1)^j\gamma^{2j}. \end{aligned} \tag{7.16}$$

According to the second formula of the system (6.34) and Formulae (6.36):

$$\gamma = -\tau_q l + \frac{1}{3!}\tau_{3q}l^3 - \frac{1}{5!}\tau_{5q}l^5 + \cdots \tag{7.17}$$

The formulae for determining the derivatives of the function τ are presented in Appendix A.

Substituting polynomials of functions $\gamma, \gamma^2, \gamma^3 \ldots$ in powers of l into Eq. (7.16) yields polynomials for the functions $\sin\gamma \cos\gamma$ of the form:

$$\sin\gamma = \sum_0^3 S_j l^{2j+1}, \tag{7.18}$$

$$\cos\gamma = \sum_0^3 C_j l^{2j}, \tag{7.19}$$

where S_j, C_j ($j = 0, 1 \ldots 3$) are functions of latitude that depend on the derivatives of the function τ and are equal to:

$$
\begin{aligned}
S_0 &= -\tau_q \\
S_1 &= \tfrac{1}{3!}\left(\tau_q^3 + \tau_{3q}\right) \\
S_2 &= \tfrac{1}{5!}\left(-\tau_{5q} - 10\tau_q^2\tau_{3q} - \tau_q^5\right) \\
S_3 &= \tfrac{1}{7!}\left(\tau_{7q} + 70\tau_q\tau_{3q}^2 + 21\tau_q^2\tau_{5q} + 35\tau_q^4\tau_{3q} + \tau_q^7\right),
\end{aligned}
\tag{7.20}
$$

$$
\begin{aligned}
C_0 &= -\tfrac{1}{2!}\tau_q^2 \\
C_1 &= \tfrac{1}{4!}\left(4\tau_q\tau_{3q} + \tau_q^4\right) \\
C_2 &= \tfrac{1}{6!}\left(-10\tau_{3q}^2 - 6\tau_q\tau_{5q} - 20\tau_q^3\tau_{3q} - \tau_q^6\right) \\
C_3 &= \tfrac{1}{8!}\left(8\tau_q\tau_{7q} + 56\tau_{3q}\tau_{5q} + 280\tau_q^2\tau_{3q}^2 + 56\tau_q^3\tau_{5q} + 56\tau_q^5\tau_{3q} + \tau_q^8\right)
\end{aligned}
\quad . \tag{7.21}
$$

Since the functions m and γ of the best conformal projection are equal to the functions $(a + b - 1)$ and β of the ideal Airy projection, Formulae (7.18–7.21) can be used to find the functions $r \sin \beta$, $r \cos \beta$ of the Airy ideal projection:

$$
\begin{aligned}
r \sin \beta &= r \sum_0^J S_j l^{2j+1} \\
r \cos \beta &= r \sum_0^3 C_j l^{2j},
\end{aligned}
\tag{7.22}
$$

and the functions $r(a + b - 1)\sin \beta$ and $r(a + b - 1)\cos \beta$ of an ideal Airy projection can be defined using Eq. (7.15), namely:

$$
\begin{aligned}
r(a + b - 1)\cos \beta &= m_0\left(r - \tfrac{1}{2!}r_{qq}l^2 + \tfrac{1}{4!}r_{4q}l^4 + \cdots\right) \\
r(a + b - 1)\sin \beta &= m_0\left(-r_q l + \tfrac{1}{3!}r_{3q}l^3 - \tfrac{1}{5!}r_{5q}l^5 + \cdots\right).
\end{aligned}
\tag{7.23}
$$

Substituting (7.7 7.8) and (7.22–7.23) into (7.2–7.3) and using the method of undetermined coefficients with respect to expressions of the same degree in l yields the following formulae for determining the functions $N_0', N_2', \ldots E_1', E_3' \ldots$

$$N_{0q}' + E_1' = r(m_0 + 1), \tag{7.24}$$

$$2N_2' - E_{1q}' = -r_q m_0 + rS_1, \tag{7.25}$$

$$N_{2q}' + 3E_3' = -\frac{1}{2!}r_{qq}m_0 + rC_1, \tag{7.26}$$

$$4N_4' - E_{3q}' = \frac{1}{3!}r_{3q}m_0 + rS_2, \tag{7.27}$$

$$N_{4q}' + 5E_5' = \frac{1}{4!}r_{4q}m_0 + rC_2, \tag{7.28}$$

$$6N'_6 - E'_{5q} = -\frac{1}{5!} r_{5q} m_0 + r S_3, \tag{7.29}$$

$$N'_{6q} + 7E'_7 = -\frac{1}{6!} r_{6q} m_0 + r C_3. \tag{7.30}$$

The formulae for determining the functions $N'_0, N'_2, \ldots E'_1, E'_3 \ldots$ are structured in such a way that each subsequent function can be derived from the previous one.

If we limit ourselves to terms of no higher than degree 7, we obtain a system of 7 equations with 8 unknown functions: $N'_0, N'_2, \ldots E'_1, E'_3 \ldots$ and an unknown constant: m_0. To fully define this system, we use the method of analogies.

As shown in Sect. 7.6.1, the best Airy direct cylindrical projection is a projection where one of the extremal scale factors is constant everywhere in the mapping area, i.e.:

$$\text{either } a = 1, b \leq 1, \text{ or } \; b = 1, a \geq 1. \tag{7.31}$$

In this case, the maximum linear distortion achieved at the boundary is twice as large in absolute value as the minimum distortion at the central point (see equalities (6.16–6.17)), i.e., at the boundary:

$$a_{\max} = 1 + k', \tag{7.32}$$

in the center:

$$b_{\min} = 1 - \frac{k'}{2}. \tag{7.33}$$

Obviously, following the principle of analogies, the properties of the Airy ideal direct projection should also apply to the transverse projection. According to the first formula of the system (7.23) at the boundary, specifically at the intersection of the equator and the boundary meridian:

$$r(a + b - 1)\cos\beta|_{\substack{\varphi=0,\\ l=l_0}} = m_0\left(r|_{\varphi=0} - \frac{1}{2!} r_{qq}|_{\varphi=0} l_0^2 + \frac{1}{4!} r_{4q}|_{\varphi=0} l_0^4 - \cdots\right). \tag{7.34}$$

The substitution of the formulae from Appendix A (A.6–A.10) into (7.34), taking into account (7.31–7.32), gives:

$$r(1 + k') = m_0\left(1 + \frac{1}{2!}(1 + e'^2) l_0^2 + \frac{1}{4!}\left(1 + e'^2\right)^2 (5 + 4e'^2) l_0^4 + \cdots\right). \tag{7.35}$$

At the central point, i.e., at the intersection of the equator and the central meridian, the same function is equal to:

$$r(a+b-1)\cos\beta|_{\substack{\varphi=0,\\ l=0}} = m_0 r|_{\varphi=0} = \left(1-\frac{k'}{2}\right). \tag{7.36}$$

Combining equalities (7.35) and (7.36) after identical transformations, we obtain the following formulae for finding the parameters m_0, k':

$$m_0 = \frac{3}{2+\Theta},\ k' = 2\frac{\Theta-1}{\Theta+2}, \tag{7.37}$$

where

$$\Theta = (1 + \frac{1}{2!}(1+e'^2)l_0^2 + \frac{1}{4!}(1+e'^2)^2(5+4e'^2)l_0^4 + \cdots). \tag{7.38}$$

The numerical values of the parameters m_0, k', for the WGS84 ellipsoid are presented in Table 7.1. This table presents two values of the Airy criterion, which are obtained numerically using Simpson's formula (Atkinson 1989; Dwight 1961, p. 221). The first value, Ξ_A, is defined for a doubly connected region. The second value, Ξ'_A, is determined for the six-degree zone. For this zone, calculations were performed on a grid of meridians and parallels, defined by the following conditions:

$$\begin{aligned} \varphi_k &= \varphi_{k-1} + 0.5, \varphi_{\min} = 0, \varphi_{\max} = 90 \\ l_j &= l_{j-1} + 0.5, l_{\min} = 0, l_{\max} = 3. \end{aligned} \tag{7.39}$$

Since the six-degree zone is symmetrical with respect to both the equator and the central meridian, the minimum values of $\varphi_{\min},_{\min}$ are equal to zero.

In the area limited by the curves defined by equalities (7.1), integration was carried out using the same scheme but only for those points on each parallel where the longitudes satisfied the inequality:

$$\ln m = -k_{G-K} - \frac{1}{2!}\tau_{qq}l^2 + \frac{1}{4!}\tau_{qqqq}l^4 + \ldots \leq k_{G-K}. \tag{7.40}$$

Using the second formula of system (2.5) and conditions (7.31 and 7.33), we find the value of the function $r(a-b)$ on the central meridian of the zone:

$$r(a-b)|_{l=0} = abs\left(N'_{0q} - E'_1\right) = \frac{1}{2}rk'. \tag{7.41}$$

Condition (7.41) suggests two solutions:

$$1.\quad N'_{0q} - E'_1 = \frac{1}{2}rk', \tag{7.42}$$

$$2.\quad N'_{0q} - E'_1 = -\frac{1}{2}rk'. \tag{7.43}$$

Table 7.1 Parameters of transverse projections: ideal according to Airy, Gauss–Kruger with minimum Airy criterion, UTM, Cassini–Soldner, the best from a set of close-to-equal-area

Parameter	Projection				
	Ideal according to Airy	Gauss–Kruger with a minimum criterion Airy	UTM	Cassini–Soldner	Best, close-to-equal-area
Θ	1.0013816	–	–	–	–
k'	9.2065E−04	–	–	–	–
m_0	9.9954E−01	9.9954E−01	0.9996	1	9.9954E−01
max $V_a(\%)$	0.092	0.092	0.098	0.138	0.092
min $V_b(\%)$	– 0.046	– 0.046	– 0.040	0	– 0.092
max $V_p(\%)$	0.092	0.184	0.196	0.138	0.000
min $V_p(\%)$	– 0.046	– 0.092	– 0.080	0	– 0.000
ω_{max}(arc min)	3.163	0	0	4.746	6.330
ω_{min}	0	0	0	0	0
$\Xi_{Ch}(\%)$	0.092	0.092	0.098	0.138	0.092
$\Xi_A(\%)$	0.044	0.062	0.063	0.288	0.062
$\Xi'_A(\%)$	0.038	0.054	0.052	0.050	0.052

Skipping the solution described by Eq. (7.42) as it does not lead to a projection with the minimum Airy criterion, let us move on to the solution satisfying Eq. (7.43). Solving Eqs. (7.43) and (7.24), we obtain:

$$E'_1 = r, \tag{7.44}$$

$$N'_{0q} = rm_0. \tag{7.45}$$

Integrating Eq. (7.45), we have:

$$N'_0 = m_0 \int_0^q rdq = m_0 S, \tag{7.46}$$

where S is the length of the meridian arc from the equator to a point with latitude q, which is determined using the formulae in Appendix E.

Differentiating the function E'_1 (Eq. 7.44) and substituting the resulting derivative into Eq. (7.25), while accounting for (7.20), yields the following after identical transformations:

$$N'_2 = -\frac{1}{2} m_0 r_q. \tag{7.47}$$

Similarly, by differentiating (7.47) and substituting the resulting derivative into (7.26), after performing identical transformations and considering (7.21), the following result is obtained:

$$E'_3 = -\frac{1}{3!} r \tau_q^2. \tag{7.48}$$

By consistently repeating the above procedure for all Eqs. (7.25–7.30) and taking into account (7.20–7.21), after the same transformations, we obtain:

$$N'_4 = \frac{1}{4!} m_0 r_{q3} + \frac{1}{4!} r\left(\tau_{q3} - 2\tau_q \tau_{qq}\right), \tag{7.49}$$

$$E'_5 = \frac{1}{5!} r\left(\tau_q^4 + 2\tau_q^2 \tau_{qq} - \tau_{q4} + 2\tau_{qq}^2 + 5\tau_q \tau_{q3}\right), \tag{7.50}$$

$$N'_6 = -\frac{1}{6!} m_0 r_{q5} + \frac{1}{6!} r\left(6\tau_q^3 \tau_{qq} + 6\tau_q \tau_{qq}^2 - 3\tau_q^2 \tau_{q3} - 2\tau_{q5} + 9\tau_{qq} \tau_{q3} + 4\tau_q \tau_{q4}\right), \tag{7.51}$$

$$E'_7 = \frac{1}{7!} r\left(-\tau_q^6 - 6\tau_q^4 \tau_{qq} - 24\tau_q^2 \tau_{qq}^2 - 23\tau_q^3 \tau_{q3} - 15\tau_q \tau_{qq} \tau_{q3} - 6\tau_{qq}^3 - \tau_q^2 \tau_{q4} + 2\tau_{q6} - 19\tau_{q3}^2 - 13\tau_{qq} \tau_{q4} - 8\tau_q \tau_{q5}\right). \tag{7.52}$$

The formulae for calculating the mapping functions and their derivatives of the ideal projection according to the Airy criterion, which are converted for calculations by geodetic latitude, are presented in Appendix F.

The comparison between the mapping function N' of the ideal projection according to the Airy criterion and the corresponding Gauss–Kruger mapping function shows that the first two terms of these functions coincide (see Eqs. 6.29, 7.46–7.47). Additionally, the comparison of such a parameter as the linear scale factor along the central meridian, the value of m_0 (see Table 7.1), for the ideal projection with the corresponding parameter of conformal projections (Gauss–Kruger and UTM), shows that this value completely coincides with the similar value of the best conformal projection according to Airy and differs by 6E–05 from the corresponding value of the UTM projection. All the abovementioned results allow us to conclude that the mapping function N' of the Airy ideal projection differs from the corresponding function of the abovementioned conformal projections in terms of quantities of the fourth order of smallness, assuming that the quantity l is of the first order of smallness. This difference in these coordinates is absent on the central meridian and increases with distance from it.

Notably, the main difference between Airy's ideal projection and the best Airy conformal projection is as follows. The first is the best according to Airy among all possible projections for a doubly connected mapping area; the second is the best according to Airy for the same area among conformal projections. Figure 7.2a shows a graph of changes in extremal linear distortions along the equator of the obtained

Airy ideal projection. The extremal distortions of this projection were calculated using the derivatives of the mapping functions presented in Appendix F.

The comparison of the extremal linear distortion graphs in Figs. 7.2a and 6.3a allows us to conclude that the constructed projection is ideal according to the Airy criterion for a doubly connected region bounded by two curves, which serve as parallels for transverse projections.

Figure 7.3 shows symbol-charts of linear distortions plotted on the map grids of the Gauss–Kruger projection and the ideal Airy projection for the six-degree zone. The linear distortions of the ideal projection were calculated using the standard Formulae (1.74, 1.134). The same equations were applied to determine the linear distortions of the Gauss–Kruger projection.

The cartographic grids in Fig. 7.3 are fragmentary, comprising four fragments. These fragments are situated in belts with the following latitudinal ranges:

I. $0 \leq \varphi \leq 2$;
II. $30 \leq \varphi \leq 32$;
III. $60 \leq \varphi \leq 62$;
IV. $87 \leq \varphi \leq 90$.

Considering that the six-degree zone is difficult to display clearly in a single image, fragmenting the cartographic grid is the most optimal solution. Comparing these fragments allows for an analysis of the changes in linear distortions as one moves from fragment to fragment.

As expected, for the Gauss–Kruger projection, the symbol-charts of linear distortions are circles, with radii proportional to the magnitude of the distortions, and the color indicates the sign of the distortion.

For an ideal projection, which is equidistant, the linear distortion symbol-charts consist of two lobes oriented along the meridians.

7.2 Comparison of the UTM Projection and Gauss–Kruger Projection with the Minimum Airy Criterion

The most common variant of the Gauss–Kruger projection is the UTM projection. The linear scale factor along the central meridian of this projection is as follows (Osborn 2013, p. 135):

$$m_{0\,\mathrm{UTM}} = 0.9996. \tag{7.53}$$

Figure 7.2b shows two graphs of extremal linear distortions for two conformal projections: the UTM and the best according to the Airy criterion. The graphs are almost identical. Notably, the second projection is determined to be the best for the doubly connected mapping area, as shown by the white dotted lines in Fig. 7.1.

The UTM projection appears to be a variant of the Gauss–Kruger projection, for which the linear scale factor on the central meridian is chosen to minimize the Airy

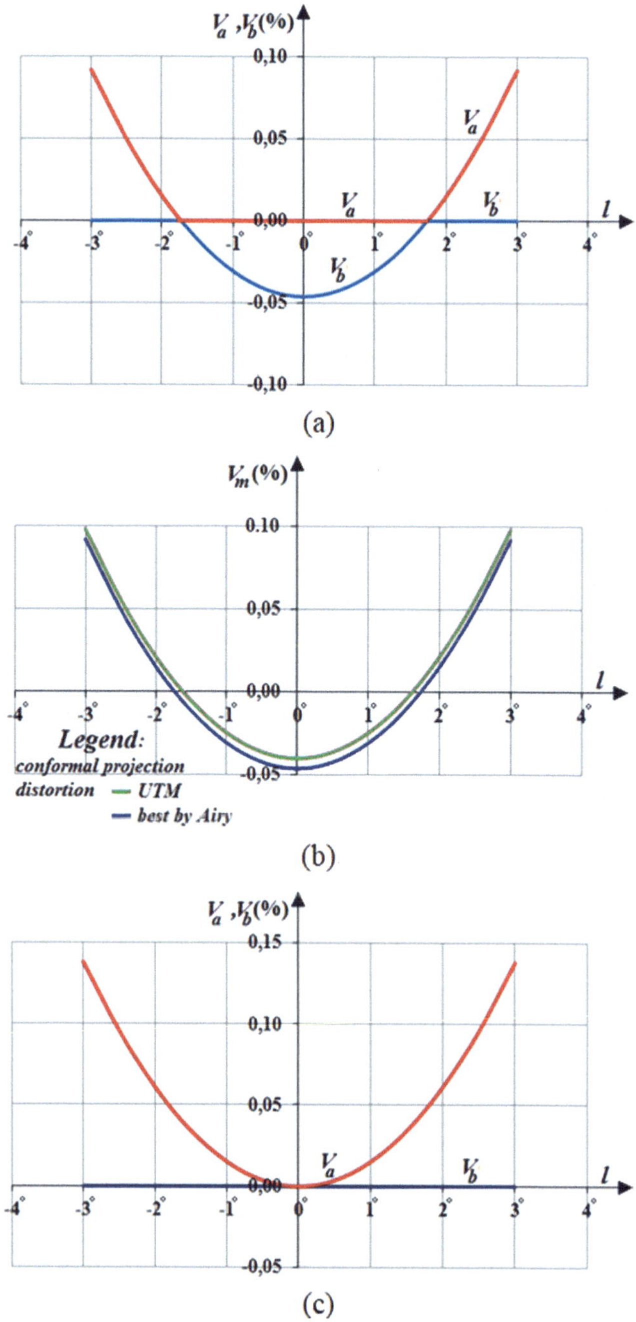

Fig. 7.2 Graphs of changes in extremal linear distortions along the equator of transverse cylindrical projections: **a** ideal according to the Airy criterion; **b** variants of conformal projections: UTM (green), best according to the Airy criterion (blue); **c** Cassini–Soldner

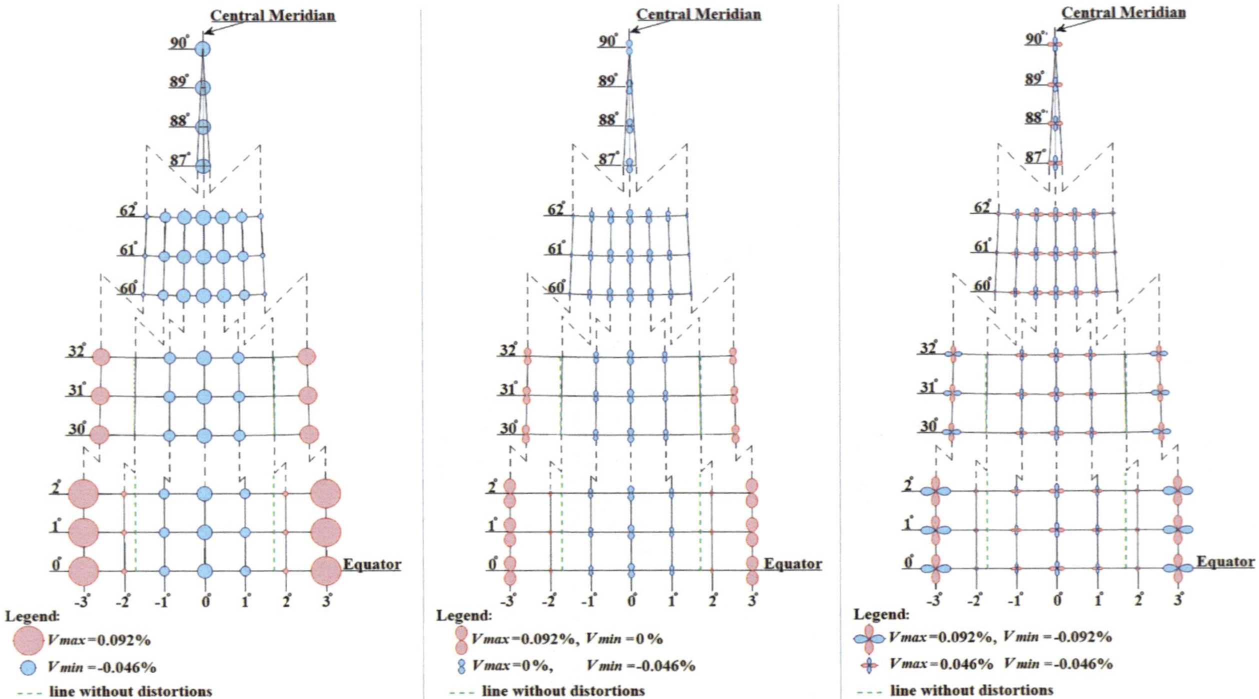

Fig. 7.3 Fragmented diagram of symbol-charts of linear distortions for: **a** Gauss–Kruger projection with a minimum value of the Airy criterion; **b** ideal projection according to the Airy criterion; **c** the best projection from a set of close-to-equal-area projections

criterion for a simply connected six-degree zone. This is evidenced by the values of the Airy criterion for this projection presented in Table 7.1. Specifically, the Airy criterion calculated for the six-degree zone for the UTM projection (0.052%) is lower than that for the conformal projection, best by the Airy criterion (0.054%). However, for the doubly connected mapping region shown in Fig. 7.1, the situation is reversed, with the Airy criterion of the UTM projection (0.063%) being higher than that of the best conformal projection according to the Airy criterion (0.062%).

7.3 Relationship Between the Ideal Airy Projection and the Gauss–Kruger Projection with Minimum Airy Criterion

The difference in coordinates ΔN_{G-K} between the mapping functions N of two projections, the ideal according to Airy and Gauss–Kruger with the minimum Airy criterion, is determined by the following formula (see Appendix F):

$$\Delta N_{G-K} = N' - N_{G-K}, \tag{7.54}$$

where the functions on the right side of the equation are determined by the first formulae of systems (6.28, 7.4) and by equalities (7.46–7.47, 7.49, 7.51).

Similarly:

$$\Delta E_{G-K} = E' - E_{G-K}, \tag{7.55}$$

where the functions E', E_{G-K} are determined by the second formulae of systems (6.28, 7.4) together with equalities (7.44, 7.48, 7.50, 7.52).

Considering $\Delta N_{G-K}, \Delta E_{G-K}$ as elements of a vector in the plane of an ideal Airy projection, we find its length and orientation:

$$S = \sqrt{\Delta N^2 + \Delta E^2}, \ \tan = \frac{\Delta E}{\Delta N}. \tag{7.56}$$

Figure 7.4 shows the displacement vectors of the projection points during the transition from the Gauss–Kruger projection to the ideal projection.

Since the vectors practically coincide with the images of parallels, for clarity, they are calculated not for the points of intersection of integer values of meridians and parallels but for points with geodetic coordinates:

$$l = \pm 0.5^\circ, \pm 1.5^\circ, \pm 2.5^\circ, \varphi = 0.5^\circ, 1.5^\circ, 2.5^\circ.$$

The maximum displacement of points occurs at the equator. Therefore, Fig. 7.5 presents a graph of the change in the ΔS value depending on the longitude along the equator.

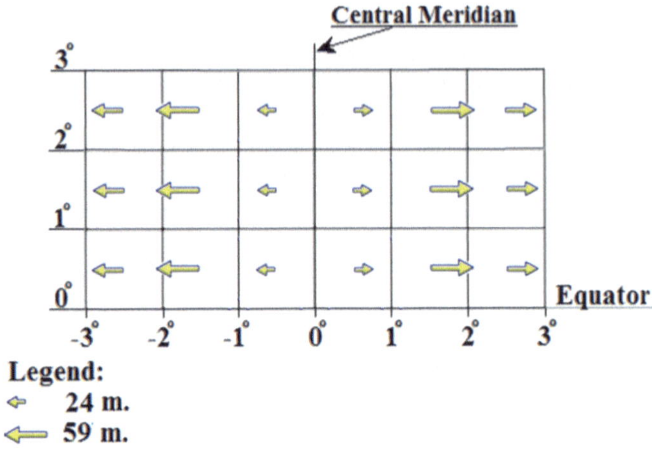

Fig. 7.4 Vector graph of the displacement of projection points during the transition from a Gauss–Kruger projection to an ideal projection

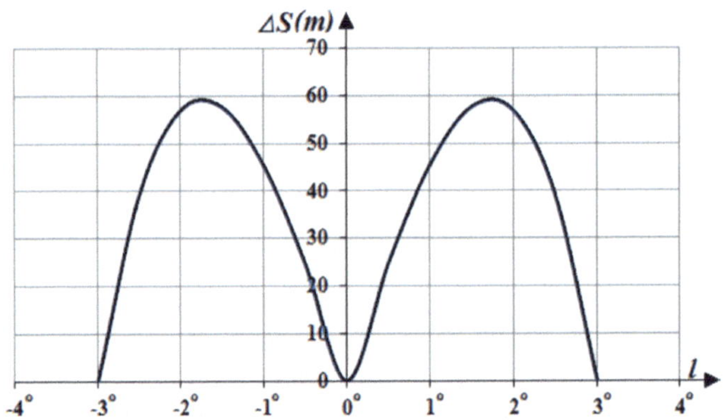

Fig. 7.5 Graph of changes in the value of ΔS for points of the Earth's ellipsoid during the transition from the Gauss–Kruger projection to the ideal Airy projection along the equator

According to Figs. 7.4 and 7.5, the points shift almost perpendicularly toward the central meridian, with the displacement value, which is zero on the central meridian, first increasing to a certain maximum and then decreasing to zero on the boundary meridians. The maximum shear value is 59.5 m. Considering that the points of the boundary meridians and the central meridian do not shift, the overall size of the zone along the E axis when moving from the Gauss–Kruger projection to the ideal Airy projection remains unchanged.

Using Formulae (7.54–7.55), the transition from the Gauss–Kruger projection to the ideal Airy projection can be made relatively simple.

7.4 Relationship Between the Ideal Projection According to the Airy Criterion and the Cassini–Soldner Projection

Since the resulting ideal projection according to the Airy criterion is equidistant, it should be similar to the Cassini–Soldner projection (Pedzich 2007; Gacoki 2018). The theoretical difference is related to the value of the linear scale factor on the central meridian. In the ideal projection, the linear scale factor is equal to a constant m_0; in the Cassini–Soldner projection, it is equal to unity. The working formulae for the Cassini–Soldner projection, as presented in (Snyder 1987, p. 92) and adapted for comparison, are as follows:

$$\begin{aligned} N_{C-S} &= S - \tfrac{l^2}{2!} r\tau_q + \tfrac{l^4}{4!} r\left(6\tau_q\tau_{qq} + \tau_q\right) \\ E_{C-S} &= rl - \tfrac{l^3}{3!} r\tau_q^2 + \tfrac{l^5}{5!}\left(8\tau_q^2\tau_{qq} + \tau_q^4\right), \end{aligned} \tag{7.57}$$

or, taking into account the equalities of Appendix A:

$$N_{C-S} = S + \frac{l^2}{2!} r \sin\varphi + \frac{l^4}{4!} r \sin\varphi \cos^2\varphi \left(5 - t^2 + 6^2\right), \tag{7.58}$$

$$E_{C-S} = rl - \frac{l^3}{3!} r \sin^2\varphi - \frac{l^5}{5!} r \sin^2\varphi \cos^2\varphi \left(8 - t^2 + 8^2\right). \tag{7.59}$$

The refined working formulae for the mapping functions and their derivatives of the Cassini–Soldner projection are presented in Appendix G.

Theoretically, the coordinate E of the ideal projection should completely coincide with the corresponding coordinate of the Cassini–Soldner projection. However, the comparison of Formulae (7.59) and (7.4), considering (7.44, 7.48, 7.50, 7.52), indicates the opposite. The difference between the formulae is as follows:

$$\Delta E_{C-S} = E' - E_{C-S} = r\frac{l^5}{30} e'^2 \cos^4\varphi \cos 2\varphi - r\frac{l^7 \cos^6\varphi}{5040} t^2\left(136 - 88t^2 + t^4\right). \tag{7.60}$$

The difference between the mapping functions along the E axis of the ideal projection and the Cassini–Soldner projection may indicate that the Cassini–Soldner projection formulae are more approximate compared to the Gauss–Kruger projection formulae.

This approximation has virtually no effect on distortions within the six-degree meridian zone, as evidenced by the value of the Airy criterion for the zone, which is 0.050% (see Table 7.1). However, errors in the formulae affected distortions outside the zone within the mapping area bounded by the curves described by Eq. (7.1). According to Table 7.1, the value of the Airy criterion for the Cassini–Soldner projection in this region is quite large and is equal to 0.288%.

Figure 7.6 shows a vector graph of the displacement of the projection points during the transition from the Cassini–Soldner projection to the ideal one. Figure 7.7 shows a graph of changes in the same value along the central meridian.

According to Figs. 7.6 and 7.7, during the transition from the Cassini–Soldner projection to the ideal projection, the points shift linearly along the N-axis. The farther the distance from the equator, the greater the displacement. The pole points have a maximum displacement of 4604 m. Thus, in general, during the transition

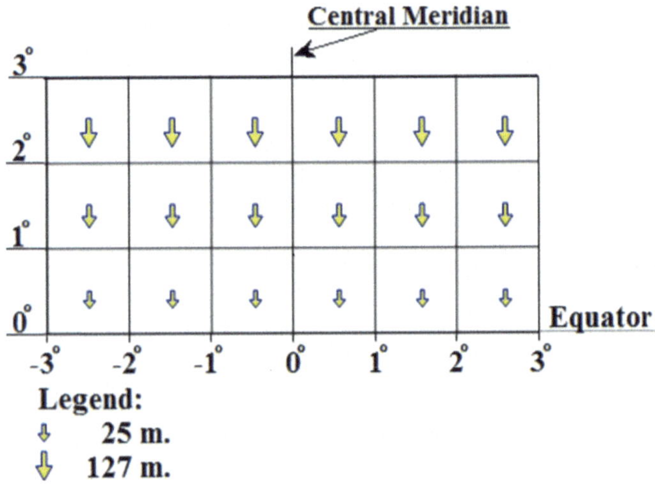

Fig. 7.6 Vector graph of the displacement of the projection points during the transition from the Cassini–Soldner projection to the ideal projection

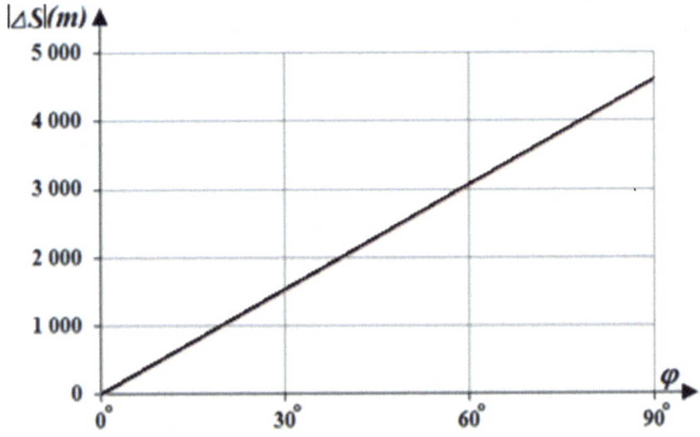

Fig. 7.7 Graph of changes in the shift of points of the Earth's ellipsoid along the central meridian during the transition from the Cassini–Soldner projection to the ideal projection

from the Cassini–Soldner projection to the ideal projection according to the Airy criterion, the plane is compressed vertically.

7.5 Projection Close-to-Equal-Area Associated with Gauss–Kruger Projection

As shown in Sect. 4.5, the best projections form a triad. In a triad containing the Gauss–Kruger projection, there is not only an ideal projection according to the Airy criterion but also the best one, which is close to equal-area. Since the mapping functions of the Gauss–Kruger and ideal projections are already known, the mapping functions of the best close-to-equal-area projection can, according to (4.56), be determined from the equalities:

$$\begin{aligned} N_E &= 2N' - N_{G-K} \\ E_E &= 2E' - E_{G-K}, \end{aligned} \tag{7.61}$$

or, using the differences between the above mapping functions (see Formulae (7.54–7.55)):

$$\begin{aligned} N_E &= N_{G-K} + 2\Delta N_{G-K} \\ E_E &= E_{G-K} + 2\Delta E_{G-K}. \end{aligned} \tag{7.62}$$

In equalities (7.61 and 7.62), N_E, E_E are the mapping functions of the best projection from the set of close-to-equal-area projections, constructed for a doubly connected narrow zone with boundary curves defined by Eq. (7.1).

Figure 7.3c shows a fragmentary diagram of symbol-charts of linear distortions for the projection under consideration. Since this projection practically does not distort the area, then the symbol-charts have four petals of equal size. In the central part of the projection, the red petals are oriented along the parallels, while the blue petals are oriented along the meridians, indicating that the linear dimensions of the objects increase along the parallels, and decrease along the meridians. At the periphery of the projection, the situation is reversed, with the red petals oriented along the meridians and the blue petals oriented along the parallels.

Table 7.1 presents the main characteristics of the projection under consideration. It is shown that the area distortion of the projection is practically equal to zero, and the maximum distortion of angles occurring at the points of the boundary meridians is 6.3 arc minutes, which is twice as large as the maximum distortion of the angles in the ideal projection from the triad (3.2 arc minutes). The maximum linear distortion is 0.09%, which is the same as the maximum linear distortion of the other projections in the triad.

7.6 Accounting for Directional Distortions in Triad Projections

The main advantage of the ideal projection according to the Airy criterion, and the best among the set of close-to-equal-area projections, is the reduction of area distortions. In an ideal projection according to the Airy criterion, area distortions are reduced by half compared with the area distortions of a conformal projection. In a projection close to equal-area, area distortions are practically equal to zero (see Table 7.1). At the same time, these projections introduce distortions in directions and angles.

As mentioned above in Sect. 7.6, when working with topographic projections, various corrections are introduced to the directions. Only one of the listed corrections (the correction for the transition from normal sections on the surface of the ellipsoid to geodesic lines) is unrelated to the projection. The other corrections depend on the properties of the projection used. Let us examine these corrections in more detail.

7.6.1 Correction for the Transition from the Azimuth of Geodesics on the Ellipsoid to the Azimuth of Their Image on the Plane

The azimuths are not distorted when transitioning from an ellipsoid to a plane in conformal projections. Therefore, let us consider azimuth corrections for the remaining two projections of the triad. For all continuously differentiable projections, this correction can be calculated using Formula (1.30). For the Airy ideal projection under consideration, the approximate formula takes the following form:

$$\delta_{\tilde{\alpha}} \cong \mathrm{DG}_0 + \mathrm{DG}_1 l + \mathrm{DG}_2 l^2 + \mathrm{DG}_3 l^3, \tag{7.63}$$

where DG_0, DG_1, DG_2, DG_3 are functions of latitude and azimuth on the ellipsoid and are equal to:

$$\begin{aligned}
\mathrm{DG}_0 &= d_0 \sin\alpha\cos\alpha \\
\mathrm{DG}_1 &= d_1 \sin\varphi \sin^2\alpha \\
\mathrm{DG}_2 &= -(\tfrac{1}{2}(1+{}^2)\cos^2\varphi + d_2 \sin^2\varphi)\sin\alpha\cos\alpha \\
\mathrm{DG}_3 &= -\big(\tfrac{1}{3}(1+{}^2)(2+4^2+d_{3c})\cos^2\varphi + \tfrac{2}{3}d_{3s}\sin^2\varphi\big)\sin\varphi\sin^2\alpha.
\end{aligned} \tag{7.64}$$

The formulae for determining the constants $d_0, d_1, d_2, d_{3c}, d_{3s}$ and their numerical values are presented in Table 7.2.

The correction $\delta_{\tilde{\alpha}}$ calculated using Formula (7.63) is expressed in radians. The root-mean-square error of the formula is 0.04 arc sec (1.9E−07 radians). The correction to azimuths $\delta_{\tilde{\alpha}}$ obtained using Formula (7.63) was compared with the correction

Table 7.2 Parameters of an ideal Airy projection depending on the parameter m_0

Parameter	Equation	Numerical value
m_0	–	9.99539673947239E−01
d_0	$2\frac{(1-m_0)}{(1+m_0)}$	4.60432027190182E−04
d_1	$2\frac{(1-m_0)}{m_0}$	9.2107610084852E−04
d_2	$4\frac{(1-m_0)}{m_0(1+m_0)}$	9.21288147316675E−04
d_{3c}	$\frac{1}{m_0}$	1.00046053805042
d_{3s}	$\frac{(1-m_0)(3-m_0)}{m_0^2}$	9.21712386736186E−04
Γ_1	$\frac{(1-m_0^2)}{m_0^2}$	9.21288196143772E−04

calculated using the exact Formula (1.30). The calculations were carried out on a grid defined by equalities (7.39). The mean square error was determined using the following equation:

$$\theta_{\delta\alpha} = \sqrt{\frac{\sum_0^J (\delta_{\tilde{\alpha}} - \delta_\alpha)^2}{i_{\max} j_{\max} k_{\max} - 1,}} \tag{7.65}$$

where δ_α is the correction to the azimuth on the ellipsoid, which is calculated using the exact Formula (1.30); $i_{\max}$ is the number of points inside the six-degree zone on each meridian; and $j_{\max}$ is the number of points inside the six-degree zone on each parallel. Thus, within the six-degree zone, a grid of $i_{\max} * j_{\max}$ points is constructed, where corrections are calculated in $k_{\max}$ directions at each point. Therefore, the total number of corrections involved in the calculations is as follows:

$$J = i_{\max} \cdot j_{\max} \cdot k_{\max}. \tag{7.66}$$

For a grid formed from conditions (7.39):

$$i_{\max} = 181, \quad j_{\max} = 13. \tag{7.67}$$

Since a complete circle is divided into 32 parts in such a way that:

$$\alpha_k = \frac{2\pi}{32} k = 0.19635\,\text{k(radian)},$$

then

$$k_{\max} = 33.$$

Thus,

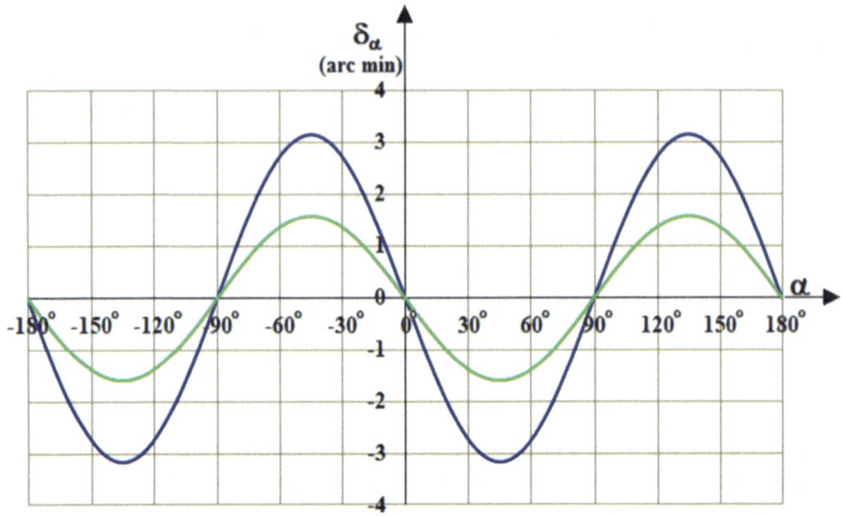

Fig. 7.8 Graph of the correction for the transition from the azimuth on an ellipsoid to the azimuth on a plane at a point with coordinates $\varphi = 0, = -3$ for the ideal projection according to the Airy criterion (green) and the best projection from the set of close-to-equal-area (blue)

$$J = 77,649.$$

Figure 7.8 shows a graph of changes in azimuth distortions, which are calculated using Formula (7.63), at the point of ideal projection with geodetic coordinates $\varphi = 0, = -3$. The graph is shown in green.

According to Fig. 7.8, the extreme value of the azimuth correction is $\pm$ 1.582 (arc min), which is expected to be half the maximum value of angular distortion of the ideal projection 3.163 (arc min) (see Table 7.1). Figure 7.8 also shows a graph of changes in the same correction but for a projection from the set of close-to-equal-area projections. The graph is shown in blue. As can be seen from the graph in Fig. 7.8, the azimuth correction when transitioning from the surface of an ellipsoid to a plane in a projection of a set of close-to-equal-area ones is almost twice as large as the corresponding correction in the ideal projection. Unfortunately, the approximate formula for a projection from the set of close-to-equal-area projections has an error that is twice as large as the corresponding formula for the ideal projection. Therefore, the correction was calculated using the exact Formula (1.30).

Figure 7.9 shows symbol-charts of azimuth distortions constructed on the grid of the ideal Airy projection (a) and the best projection of a set of close-to-equal-area projections (b).

As shown in Fig. 7.9, the azimuth correction when transitioning from the surface to the plane for the best projection from the set of close-to-equal-area projections at all points (not only at the point with coordinates $\varphi = 0, l = -3$) is twice as large as the corresponding correction for the ideal projection.

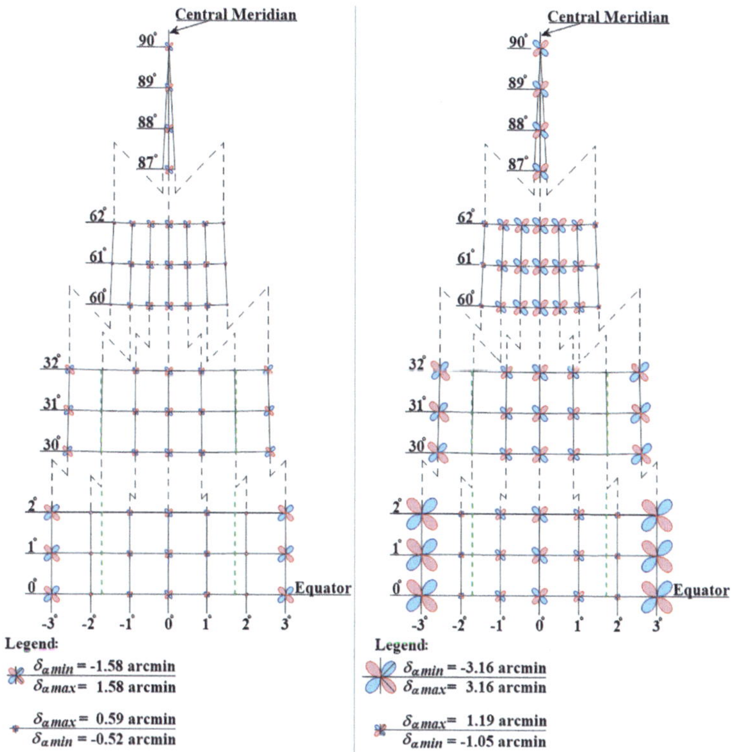

Fig. 7.9 Symbol-sharts of azimuth distortions constructed for four fragments of the cartographic grid: **a** the ideal projection according to the Airy criterion; **b** the best projection from the set of close-to-equal-area projections

7.6.2 Correction for the Transition from the Image of the Azimuth of the Geodesic Line in Projection to the Grid Azimuth

This correction is called grid convergence and is calculated by using the Eq. (1.23). According to this equation, the correction does not depend on the direction and is equal to the angle between the projected meridian and the N-axis. For the Gauss–Kruger projection, the grid convergence of meridians γ can be obtained using Formula (7.17) or by considering the formulae in Appendix A (Bagratuni 1962; Osborn 2013; Zhiping Lu et al. 2014):

$$\gamma_{G-K} = l \sin\varphi + \frac{l^3}{3}\sin\varphi\cos^2\varphi\left(1 + 3^2 + 2^4\right) + \frac{l^5}{15}\sin\varphi\cos^4\varphi\left(2 - t^2\right). \quad (7.68)$$

For an ideal projection according to the Airy criterion, the grid convergence of meridians can be obtained using the following formula:

$$\gamma_{\text{ideal}} = \frac{1}{m_0}\Bigg(-\tau_q l + \frac{l^3}{3!}\left(-\tau_q\tau_{qq} + 2\Gamma_1\tau_q^3\right) + \frac{l^5}{5!}\left(4\tau_q^3\tau_{qq} + 6\tau_q\tau_{4q} - 11\tau_q\tau_{qq}^2 + 8\tau_q^2\tau_{3q} + \tau_{5q} - 9\tau_{qq}\tau_{3q}\right)\Bigg), \tag{7.69}$$

where Γ_1 is a constant; its numerical value and the formula for calculation are presented in Table 7.2.

Substituting the values of the derivatives of the function τ from Appendix A into Eq. (7.69) yields:

$$\gamma_{\text{ideal}} = \frac{1}{m_0}\Bigg(l\sin\varphi - \frac{l^3}{3!}\sin\varphi\cos^2\varphi\left(1+{}^2+2_1t^2\right) + \frac{l^5}{5!}\sin\varphi\cos^2\varphi\left(1+{}^2\right)\left(1+52t^2\right)\Bigg). \tag{7.70}$$

Formula (7.70) was obtained using the procedure that F.N. Krasovsky used to define Formula (7.68) (Bagratuni 1962). An approximate function for calculating the grid convergence of meridians in the best projection from the set of close-to-equal-area projections can be written as follows:

$$\gamma_{E-A} = \sum_{j=0}^{3} G_{2j+1} l^{2j+1}, \tag{7.71}$$

where

$$G_1 = \tau_q\left(1 - \frac{2}{m_0}\right),\ G_3 = -\frac{1}{3!}\tau_{3q} + \frac{1}{3!m_0}\left(-2\tau_q\tau_{qq} + 8\tau_q^3\right) - \frac{4}{m_0^2}\tau_q^3 + \frac{8}{3m_0^3}\tau_q^3,$$

$$\begin{aligned}G_5 = {} & \frac{1}{5!m_0}\left(18\tau_q\tau_{qq}^2 + 48\tau_q^3\tau_{qq} - 32\tau_q^5 - 44\tau_q^2\tau_{3q} + 2\tau_{5q} - 18\tau_{qq}\tau_{3q} - 8\tau_q\tau_{4q}\right)\\ & + \frac{1}{5!}\tau_{5q} + \frac{1}{3!m_0^2}\left(-10\tau_q^3\tau_{qq} + \tau_q\tau_{4q} - 2\tau_q\tau_{qq}^2 + 3\tau_q^2\tau_{3q} + 24\tau_q^5\right)\\ & + \frac{1}{3m_0^3}\left(4\tau_q^3\tau_{qq} - 40\tau_q^5\right) + \frac{16}{m_0^4}\tau_q^5 - \frac{32}{5m_0^5}\tau_q^5.\end{aligned} \tag{7.72}$$

The graph of the difference between the grid convergence of the meridians in the ideal projection and the Gauss–Kruger projection, constructed for the boundary meridian $l = 3$, is presented in Fig. 7.10.

According to Fig. 7.10, the minimum difference between the grid convergence of the meridians of the ideal projection and the Gauss–Kruger projection, equal to – 3.15 (arc sec), is achieved at latitude $\varphi \approx 28$; the maximum difference, equal to

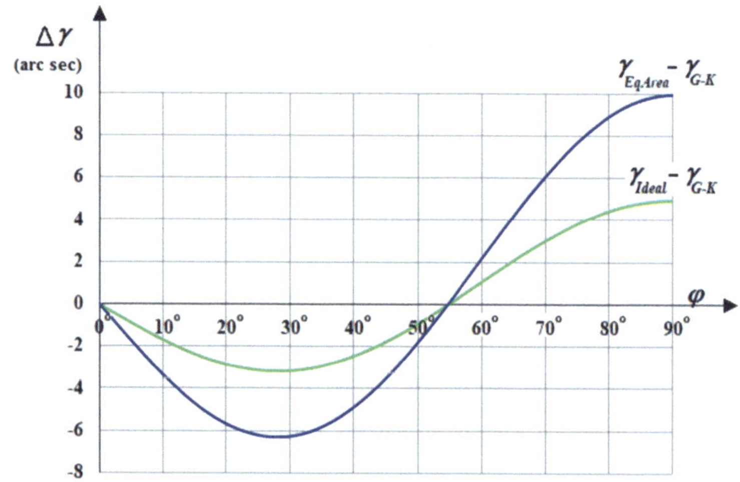

Fig. 7.10 Graph of the difference in grid convergence of meridians along the boundary meridian $l = 3$ for the following projections: ideal according to Airy and Gauss–Kruger (green); the best of the set of close-to-equal-area projections and Gauss–Kruger (blue)

$+ 4.96$ (arc sec), is achieved at the pole at $\varphi = 90$. A similar situation is observed for the difference in γ between the best projection from the set of close-to-equal-area projections and the Gauss–Kruger projection. At the minimum point, at $\varphi \approx 28$, the difference in the γ values is equal to $- 6.30$ (arc sec); at the maximum point, at $\varphi = 90$, the difference reaches $+ 9.93$ (arc sec).

7.6.3 *Correction for the Curvature of the Image of Geodesics*

The third correction applied to angles is the correction for the curvature of the image of the geodesic line (see Fig. 1.5), known as the arc-to-chord correction (Buzila et al. 2014).

The geodesic line constructed on the surface of an ellipsoid is depicted on a plane in projection as a curved line that does not coincide with the chord. Only the gnomonic projection has the property of mapping geodesics into geodesics, i.e., into chords (Karney 2013, 2024). Conformal projections do not possess this property. Therefore, when conformal projections are used to transfer information from a surface to a plane, corrections must be introduced to account for directional distortions. For the Gauss–Kruger projection, as with any other projection, arc-to-chord corrections can be determined using Eqs. (1.54, 1.66–1.70). In this case, Formulae (6.28) should be used to obtain the derivatives of the mapping functions, which are then substituted into Eqs. (1.66–1.70). However, as mentioned above, the arc-to-chord corrections in the Gauss–Kruger projection can be determined by using Eq. (1.31). Since the

Gauss–Kruger projection is conformal, all angular distortions, including azimuthal distortions, are zero. Therefore, Eq. (1.31) is rewritten as follows:

$$\hat{A} = \alpha - \gamma. \tag{7.73}$$

For the Gauss–Kruger projection, grid convergence is determined by Eq. (7.17). By differentiating (7.73) under conditions (1.64, 7.17) and substituting the resulting derivatives into (1.59, 1.69), we obtain, after identical transformations:

$$\hat{A}_s = \frac{1}{r}\left(l_{q2} - \frac{1}{3!}l^3_{q4}\right)\cos\alpha + \frac{1}{r}\left(-\frac{1}{2!}l^2_{q3} + \frac{1}{4!}l^4_{q5}\right)\sin\alpha, \tag{7.74}$$

$$\hat{A}_{ss} = \frac{1}{r^2}\left[l\left(_{q3} - _{qq2}\right) - \frac{1}{3!}l^3\left(_{q5} - _{qq4}\right)\right]\cos 2\alpha + \frac{1}{r^2}\left[\frac{1}{2}_{q2} - \frac{1}{2!}l^2\left(_{q4} - _{qq3}\right)\right]\sin 2\alpha. \tag{7.75}$$

Corrections to the initial and final directional angles of the geodesic line can be calculated for the Gauss–Kruger projection by substituting Eqs. (7.74 and 7.75) into Formulae (1.54). The arc-to-chord corrections for the remaining projections of the triad were calculated using the general Formulae (1.66–1.70), taking into account the derivatives of the corresponding mapping functions.

To visually represent how $\overline{\delta}_1$ changes, symbol-charts were constructed at the nodes of the cartographic grid, which were determined by conditions (7.39) (see Fig. 7.11). The procedure for constructing a symbol-chart is similar to that described above in Sect. 1.12. The corrections were calculated for distance of $s = 50$ km. The legends of all projections contain two symbols, the first for the point with coordinates $\varphi = 0, l = -3$ and the second for the point with coordinates $\varphi = 62, l = +1$.

According to Fig. 7.11, the largest corrections occur at the points on the equator with longitudes equal to ± 3°. For the Gauss–Kruger projection, at points east of the central meridian, the maximum correction value is 42.6 arc sec, which occurs at a grid azimuth of 0°, and the minimum value is (– 42.6) arc sec at a grid azimuth of 180°. For points west of the central meridian, the situation is mirrored with respect to the parallel.

Unlike the symbol-charts of the Gauss–Kruger projection, which are symmetrical with respect to both the meridian and the parallel, the symbol-charts of the remaining projections of the triad are symmetrical only with respect to the parallel.

The extremum of corrections in the ideal projection according to the Airy criterion is not achieved along the meridians but along vectors directed outside the zone. At points at the equator with longitudes of ± 3°, the maximum value of the arc-to-chord correction is 47.7 arc sec, and the minimum is – 47.7 arc sec. A similar situation occurs for the projection that is the best from the set of close-to-equal-area projections. The maximum correction value achieved at the same points is 62.5 arc seconds, and the minimum correction value is (– 62.5) arc sec.

Figure 7.12 shows a graph of the changes in the arc-to-chord corrections $\overline{\delta}_1$ depending on the azimuth on the ellipsoid at the equator at a point with longitude

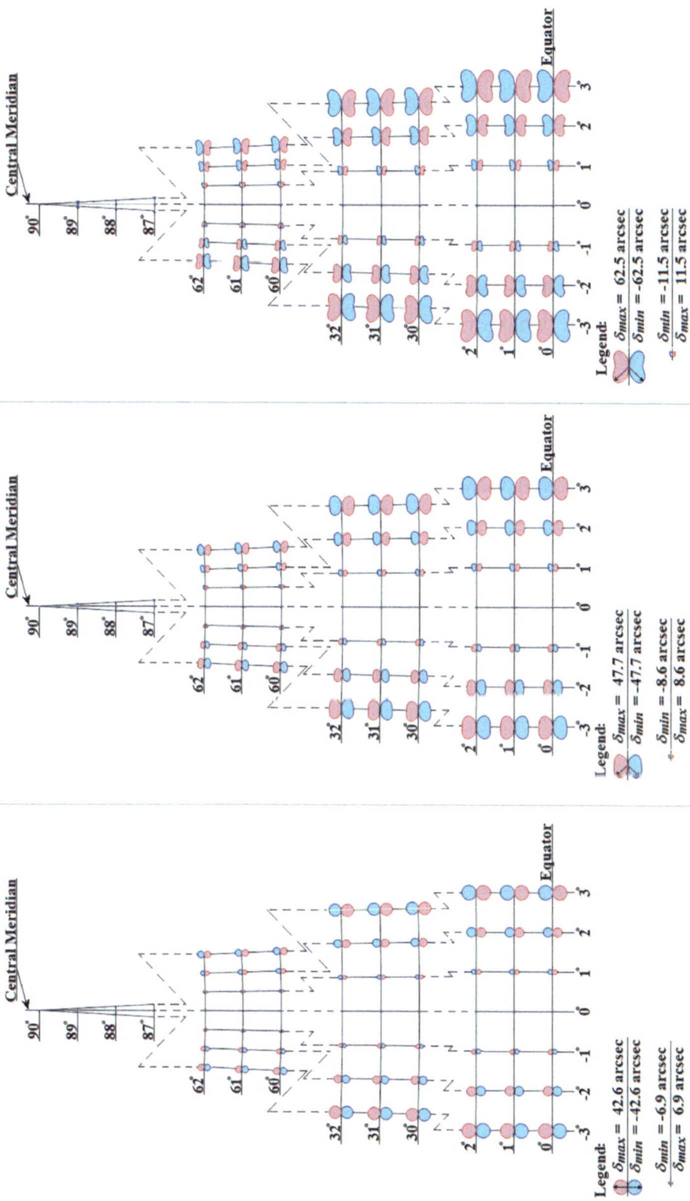

Fig. 7.11 Symbol-charts of arc-to-chord corrections $\overline{\delta}_1$ for a distance of $s = 50$ km, constructed on fragments of cartographic grids: (a) Gauss–Kruger projection; (b) ideal projection according to the Airy criterion; (c) projection, the best from the set of close-to-equal-area projections

$l = -3$. The red graph corresponds to the corrections for the Gauss–Kruger projection, the blue graph corresponds to the corrections for a projection from the set of close-to-equal-area projections, and the green graph corresponds to the corrections for the ideal projection according to the Airy criterion.

Considering the closeness of the graphs presented in Fig. 7.12, we can conclude that, to a first approximation, the corrections for all three projections are nearly equal. However, in the second approximation, differences in the corrections cannot be neglected. The graphs of the arc-to-chord corrections for the ideal projection according to the Airy criterion and the best projection from the set of close-to-equal-area projections, unlike the similar graph for the Gauss–Kruger projection, are not symmetrical with respect to the azimuth. For negative azimuths directed away from the central meridian, the extremum of the graphs is greater than that for positive azimuths directed toward the central meridian. This conclusion fully aligns with the earlier analysis of the behavior of arc-to-chord corrections on the basis of the symbol-charts in Fig. 7.11.

On the central meridian, according to Fig. 7.11, arc-to-chord corrections are vanishingly small. However, there is no single point in the map area where the curvature of the geodesics is zero in all directions. Moreover, the shapes of the symbol-charts on the central meridian and in its immediate vicinity show that at each point of the central meridian, there are not two but four directions in which there is no geodesic curvature, and the arc-to-chord corrections are equal to zero, as are the four directions in which the corrections are extremal. To illustrate this phenomenon, Fig. 7.13 separately shows symbols-charts of corrections along the equator near the central meridian. The symbol-charts are constructed for meridians close to the central meridian, namely, 0, $\pm$ 0.05°, and $\pm$ 0.10°.

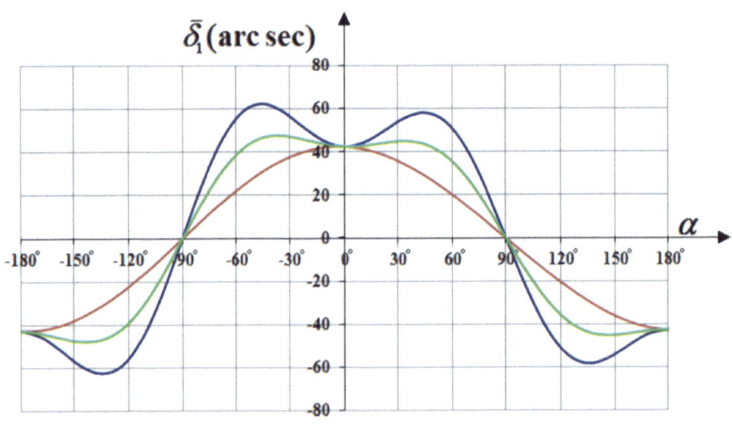

Fig. 7.12 Graphs of changes in arc-to-chord corrections at the beginning of a geodesic line 50 km long for the following projections: Gauss–Kruger (red), ideal according to the Airy criterion (green), the best from the set of close-to-equal-area projections (blue) for a point with geodetic coordinates $\varphi = 0, = -3°$

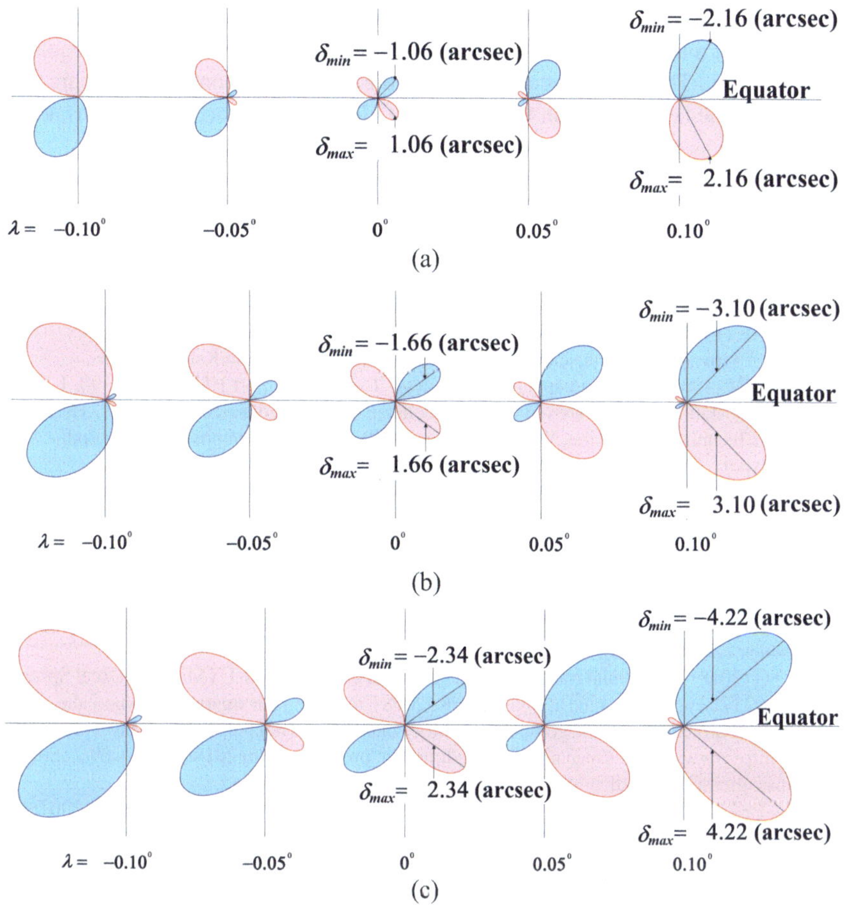

Fig. 7.13 Symbol-charts of the equator near the central meridian for the following projections: **a** Gauss–Kruger; **b** ideal according to the Airy criterion; **c** best from the set of close-to-equal-area projections

A comparison of the symbol-charts for all projections in the triad allows us to conclude that near the central meridian, they have similar shapes. However, there are some peculiarities as well. In particular, the absolute value of the corrections for the ideal projection is 1.5 times greater than the corrections for the Gauss–Kruger projection and 1.5 times less than the corrections for the best projection from the set of close-to-equal-area projections.

According to Fig. 7.13a, the extremal values of the corrections at the intersection of the equator and the central meridian for the Gauss–Kruger projection are equal to $\pm$ 1.06 arc sec. The extremal corrections for the ideal projection are $\pm$ 1.66 arc sec, and that for the best projection from the set of close-to-equal-area projections are $\pm$ 2.34 arc sec.

References

Atkinson KE (1989) An Introduction to Numerical Analysis, 2nd edn. Wiley, p 663. ISBN 0-471-50023-2. https://www.math.science.cmu.ac.th/docs/qNA2556/ref_na/Katkinson.pdf

Bagratuni GV (1962) Course in spheroidal geodesy, Technical Report FTD-MT-64-390, US Air Force, translation of Kurs sferoidicheskoi geodezii, 1967. (Geodezizdat, Moscow, 1962), by Foreign Technology Division, Wright-Patterson AFB, p 305. https://apps.dtic.mil/sti/pdfs/AD0650520.pdf

Buzila A-M, Moldovan M-O, Tibrea MD, Ursu E (2014) Comparative analysis of arc-to chord corrections between the Gauss–Kruger projected system and the Stereographic 1979 projected system for the same side of triangulation network. J Young Sci II:161–166. ISSN 2344-1283. https://journalofyoungscientist.usamv.ro/pdf/vol_II_2014/art28.pdf

Dwight HB (1961) Tables of integrals and other mathematical data. The MacMillan Company, New York. https://www.dce.uni-sofia.bg/shares/pk/Mathematics-for-Pharmacy-BG/Dwight.%20Tables%20of%20Integrals.pdf

Gacoki TG (2018) Polynomial transformation of cassini coordinates to UTM on the Excel Spreadsheet (9237). Paper presented at FIG congress 2018 Embracing our smart world where the continents connect: enhancing the geospatial maturity of societies Istanbul, Turkey, May 6–11, 2018, p 12. https://www.fig.net/resources/proceedings/fig_proceedings/fig2018/papers/ts09e/TS09E_thomas_gicira_9237.pdf

Karney CFF (2013) Algorithms for geodesics. J Geodesy 87:43–55. https://doi.org/10.1007/s00190-012-0578-z. https://link.springer.com/article/https://doi.org/10.1007/s00190-012-0578-z

Karney CFF (2024) Geodesics on an arbitrary ellipsoid of revolution, J Geodesy 98(4):14. https://link.springer.com/content/pdf/10.1007/s00190-023-01813-2.pdf

Lu Z, Qu Y, Qiao S (2014) Geodesy, Introduction to Geodetic Datum and Geodetic Systems, Springer, Berlin Heidelberg, p 417. https://doi.org/10.1007/978-3-642-41245-5. http://ndl.ethernet.edu.et/bitstream/123456789/59034/1/697.pdf

Osborn P (2013) The Mercator projections. The normal and transverse Mercator projections on the sphere and the ellipsoid with full derivations of all formulae. Edinburgh, p 254. https://doi.org/10.5281/zenodo.35392. https://zenodo.org/records/35561

Pedzich P (2007) Properties of Cassini-Soldner projection of the whole ellipsoid. Polskie Towarystwo Informacii Przestrzennej. Roczniki Geomatyki 5(2):73–84. http://rg.ptip.org.pl/index.php/rg/article/viewFile/RG2007-2-Pedzich/1038

Snyder JP (1987) Map projections—a working manual, U.S. Geological Survey professional paper 1395, USPOGO, Washington D.C, p 397. https://doi.org/10.3133/pp1395

Chapter 8
Ideal and Best Projections According to Chebyshev's Criterion

8.1 Ideal Projection by Chebyshev's Criterion

8.1.1 Ideal Projection Theorem

Consider a surface where all projections onto the plane have the following property: the maximum scale factor reaches a maximum at the boundary of the region. These surfaces include a sphere and an ellipsoid. Let us choose a geodetic polar coordinate system ξ, ϑ. The first fundamental form of the surface in this coordinate system, as well as in the isometric coordinate system q, λ (Eq. 1.39), can be written as follows (Meshcheryakov 1968):

$$ds^2 = d\xi^2 + \Omega^2 d\vartheta^2, \tag{8.1}$$

where Ω is a differentiable function that satisfies the following conditions:

$$\Omega|_{\xi=0} = 0, \quad \lim_{\Omega \to 0} (\xi/\Omega) = 1. \tag{8.2}$$

The choice of the coordinate system does not affect the property of the projection being the best according to any criterion. A projection that is ideal according to Chebyshev's criterion in one coordinate system must also be ideal in another coordinate system. Let us choose a geodetic polar coordinate system such that the origin is located inside the mapping area, and the following equality is satisfied at the boundary:

$$\frac{\xi}{\Omega} = C_1 = \text{const}. \tag{8.3}$$

Let there be a polar coordinate system ρ, δ on the plane, satisfying the following conditions:

E. Novikova, *Best Map Projections*, https://doi.org/10.1007/978-3-031-78334-0_8

$$\begin{aligned} \rho|_{\xi=0} &= 0 \\ \delta|_{\vartheta=0} &= T_0(\xi) \\ \delta|_{\vartheta=2\pi} &= T_0(\xi) + 2\pi \end{aligned}, \tag{8.4}$$

where $T_0(\xi)$ is some continuous function of ξ.

The first condition of system (8.4) corresponds to the choice of the origin of the coordinate system on the plane. The second and third conditions make it possible to avoid breaking the projection along the line $\vartheta = 2\pi$.

The linear scale factors along the coordinate lines $\xi = const$, $\vartheta = const$ can be written in the following form (Meshcheryakov 1968):

$$\begin{aligned} \mu_1^2 &= \rho_\xi^2 + \rho^2\delta_\xi^2 \\ \mu_2^2 &= \left(\rho_\vartheta^2 + \rho^2\delta_\vartheta^2\right)/\Omega^2 \end{aligned}. \tag{8.5}$$

In Eq. (8.5), $\rho_\xi, \rho_\vartheta, \delta_\xi, \delta_\vartheta$ are the first derivatives of the mapping functions ρ, δ with respect to the variables ξ, ϑ.

Consider a projection for which:

$$\rho = \xi(1 - H), \quad \delta = \vartheta, \tag{8.6}$$

where H is a constant value.

The linear scale factors in this projection are equal:

$$\begin{aligned} \mu_1 &= (1 - H) = \text{const} \\ \mu_2 &= \tfrac{\rho}{\Omega} = \tfrac{\xi(1-H)}{\Omega}. \end{aligned} \tag{8.7}$$

Since Eq. (8.3) is satisfied at the boundary, μ_2 at the boundary of the mapping area is equal to:

$$\mu_2|_{\text{on borderline}} = C_1(1 - H) = \text{const}. \tag{8.8}$$

Let us choose the constant C_1 such that at the boundary:

$$\mu_2 = (1 + H). \tag{8.9}$$

In this case:

$$C_1 = \frac{1 + H}{1 - H}. \tag{8.10}$$

According to (2.35, 8.7 and 8.9), the value of Chebyshev's criterion for this projection is equal to:

$$\Xi_{\text{Ch}} = H. \tag{8.11}$$

Let us show that for any continuously differentiable projection, the value of Chebyshev's criterion will not be less than H. To do this, we use the method of A. Markov (1895), described in (Novikova 1989).

Let us divide the entire set of continuously differentiable projections into two groups. The first group includes projections whose mapping functions satisfy the following equalities:

$$\rho = T_1(\xi, \vartheta), \quad \delta = T_2(\vartheta). \tag{8.12}$$

Similarly, the mapping functions of the projections in the second group are defined as:

$$\rho = T_1(\xi, \vartheta), \quad \delta = T_3(\xi, \vartheta). \tag{8.13}$$

Obviously, for each projection in the first group, there are infinitely many corresponding projections in the second group. These projections share the same value of the variable ρ but differ in the value of the variable δ.

Let us assume that in the first group, there is a projection for which Chebyshev's criterion is less than H. Then, for this projection, the minimum scale factor, and therefore the scale factor along the coordinate lines $\xi = \text{const}$, must be greater than the value $1 - H$, i.e.,

$$\mu_1 = \rho_\xi > (1 - H). \tag{8.14}$$

Using Lagrange's theorem on finite increments of a differentiable function (Bronstejn et al. 2020), we obtain:

$$\rho_\xi(A_2) - \frac{\rho(A_1) - \rho(A_0)}{\xi(A_1) - \xi(A_0)}, \tag{8.15}$$

where A_0 is the point at which $\xi = 0$, A_1 is a certain point on the boundary of the region, and A_2 is a point located between the points A_0, A_1 along the coordinate line $\xi = \text{const}$.

Taking into account the first formula of the system (8.4) and inequality (8.14), we obtain:

$$\rho_\xi(A_2) = \frac{\rho(A_1)}{\xi(A_1)} > (1 - H). \tag{8.16}$$

At the boundary, the maximum scale factor, and therefore the scale factor along the coordinate lines $\vartheta = \text{const}$, must be less than$1 + H$, i.e.,

$$\mu_2 = \frac{\sqrt{\rho_\vartheta^2 + \rho^2 \delta_\vartheta^2}}{\Omega} < (1 + H).$$

Hence,

$$\frac{\rho\delta_{\vartheta}}{\Omega} < (1+H). \tag{8.17}$$

At the boundary, the variable ϑ takes all values from 0 to 2π. Considering the second and third conditions of the system (8.4), we can conclude that δ also takes all values from 0 to 2π. Using Lagrange's theorem on finite increments of a differentiable function (Bronstejn et al. 2020), we can conclude that on the boundary of the region, there is at least one point A_1 at which the following equality holds:

$$\delta_{\vartheta}(A_1) = \frac{2\pi - 0}{2\pi - 0} = 1. \tag{8.18}$$

Taking into account Eq. (8.18), inequality (8.17) can be rewritten as follows:

$$\frac{\rho(A_1)}{\Omega(A_1)} < (1+H). \tag{8.19}$$

Combining inequalities (8.16 and 8.19) and considering Eqs. (8.3 and 8.10), we obtain:

$$(1-H) < \frac{\rho(A_1)}{\xi(A_1)} = \frac{\rho(A_1)\Omega(A_1)}{\Omega(A_1)\xi(A_1)} < \frac{(1+H)\Omega(A_1)}{\xi(A_1)} = (1-H). \tag{8.20}$$

The inconsistency of this inequality proves the impossibility of the existence of a projection in the first group with a value of Chebyshev's criterion less than H. Therefore, for any arbitrary projection of the first group, there is a point A_3 in the mapping area at which the following inequality holds:

$$\mu_1 = T_{1\xi}(A_3) \geq (1+H). \tag{8.21}$$

Let us determine the condition that the function μ_1 must satisfy at this point so that a projection belonging to the second group has Chebyshev's criterion less than H. Using Eqs. (8.5 and 8.13), we obtain the projection of the second group at point A_3.

$$\mu_1^2 = T_{1\xi}^2(A_3) + T_1^2(A_3)T_{3\xi}^2(A_3) < (1+H)^2,$$

or

$$T_1^2(A_3)T_{3\xi}^2(A_3) < (1+H)^2 - T_{1\xi}^2(A_3) \leq (1+H)^2 - (1+H)^2 = 0. \tag{8.22}$$

The inconsistency of the resulting inequality proves the impossibility of the existence of a projection in the second group with a Chebyshev's criterion value less than H.

The resulting ideal projection is orthogonal along the coordinate lines of the geodetic polar coordinate system ξ, ϑ. The maximum scale factor at the boundary is equal to the constant value $(1+H)$ and reaches a minimum equal to $(1-H)$ at the central point inside the mapping area.

8.1.2 Postel Projection

As previously mentioned, the ideal projection according to Chebyshev's criterion for the area bounded by the almucantar is the azimuthal Postel projection, whose mapping functions are written as follows (Bugayevskiy and Snyder 1995, p. 105):

$$\begin{aligned} N &= R(1-H)z\cos A \\ E &= R(1-H)z\sin A. \end{aligned} \quad (8.23)$$

The extremal scale factors are as follows:

$$\begin{aligned} b &= (1-H) = \text{const} \\ a &= \tfrac{z}{\sin z}(1-H). \end{aligned} \quad (8.24)$$

The constant H in this projection is determined by the condition that the maximum scale factor on the boundary is equal to:

$$a_{z_0} = \frac{z_0}{\sin z_0}(1-H) = (1+H). \quad (8.25)$$

Hence,

$$H = \frac{z_0 - \sin z_0}{z_0 + \sin z_0}. \quad (8.26)$$

Figure 8.1 shows a graph of the changes in H depending on z_0. Since the value of H is Chebyshev's criterion for a projection whose boundary zenith distance is z_0, H is represented in the graph as a percentage.

According to the graph in Fig. 8.1, the projection of the entire globe ($z_0 = 180°$) has a maximum linear distortion equal to 100%. The numerical value of the constant H is equal to:

$$H = 1.$$

The indicator of unambiguity for this projection (see Formula (1.109)) is as follows:

$$\overline{\sigma} = \frac{h}{R^2} = (1-H)^2 \sin z. \quad (8.27)$$

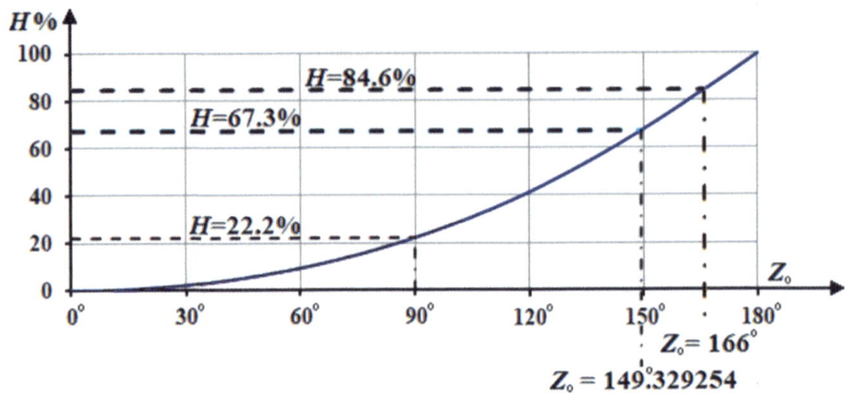

Fig. 8.1 Graph of the dependence of Chebyshev's criterion H on z_0

In the interval $0 \leq z \leq 180°$, the indicator $\overline{\sigma} = 0$. Therefore, when $H = 1$, all the projection coordinates are equal to zero, and the projection linear scale factors are also equal to zero (see Formulae (8.24–8.25)).

Consequently, constructing a Postel projection over the entire territory of the globe is impossible. Therefore, to construct the projection map grid, two boundary zenith distances were chosen: 166°, the maximum possible boundary zenith distance for the almucantar of the Airy projection at which the Airy projection remains one-to-one, and 149.329254°, the boundary zenith distance for the Airy projection, at which the Airy projection at the boundary of the mapping area has linear scale factors that are not close to zero.

Figures 8.2 and 8.3 show the Postel projection cartographic grids with Tissot's indicatrixes, built for boundary almucantars 166° and 149.329254°, respectively.

A comparison of the indicatrixes without distortion reveals that the radius of this indicatrix in Fig. 8.3 is half the size of the corresponding radius of the indicatrix in Fig. 8.2. This means that the total approximate grid size of the projection with $z_0 = 166°$ is half the size of the projection grid with $z_0 = 149.329254°$.

Table 8.1 presents the main characteristics of the Postel projection for the two boundary almucantars.

Let us show that the Postel projection is an ideal Chebyshev projection for the area limited by the almucantar.

Indeed, the Postel projection is orthogonal to the geodetic polar coordinate system z, α. The minimum scale factor is equal to the constant value $(1 - H)$, whereas the maximum scale factor at the boundary is equal to the constant value $(1 + H)$ and reaches a minimum of $(1 - H)$ at the central point inside the mapping area.

Analyzing the data in Tables 5.1 and 8.1 for projections with a boundary almucantar having a zenith distance of 149.329254°, it can be concluded that Chebyshev's criterion for the Postel projection is 119% lower than the corresponding criterion of the Airy projection.

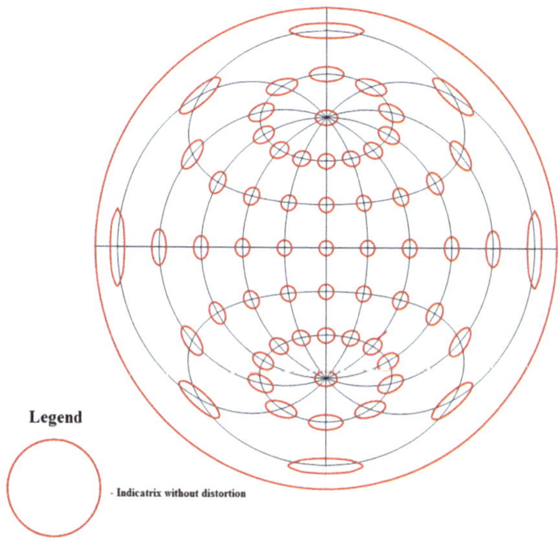

Fig. 8.2 Postel projection map grid with Tissot's indicatrixes. Meridians and parallels are drawn every 30°. The outer circle (red) is an image of the boundary almucantar with $z_0 = 166°$

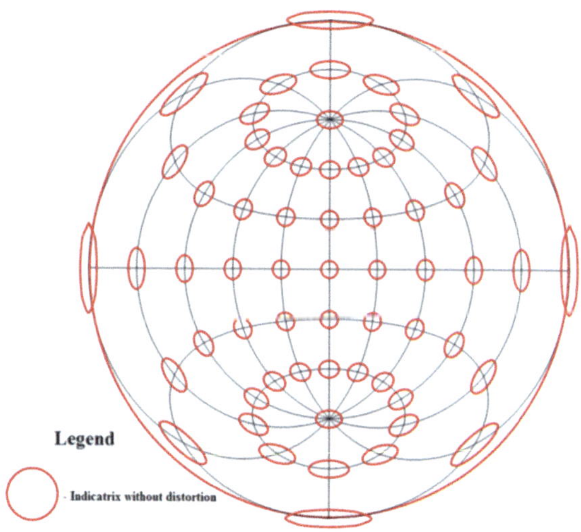

Fig. 8.3 Postel projection map grid with Tissot's indicatrixes. Meridians and parallels are drawn every 30°. The outer circle (red) represents the boundary almucantar with $z_0 = 149°.329254$

Similarly, the value of the Airy criterion for the Airy projection is 13% lower than the same value for the Postel projection. The distortion of angles in the Airy projection is 37% less than the distortion of angles in the Postel projection. Moreover, the area distortions of the Airy projection are 231% greater than those of the Postel projection.

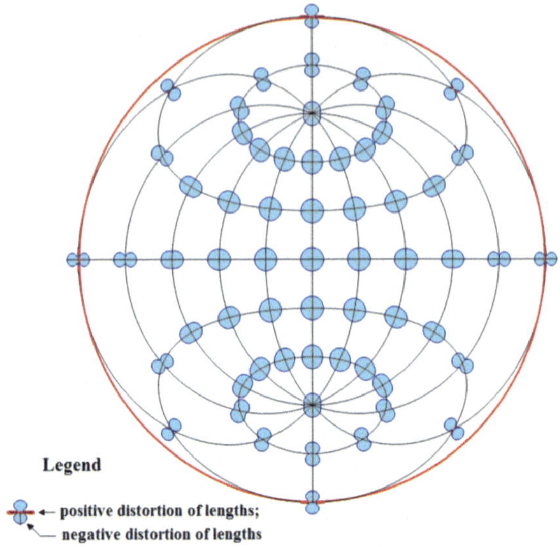

Fig. 8.4 Postel projection map grid constructed with symbol-charts of linear distortions. Meridians and parallels are drawn every 30°. The outer circle (red) represents the boundary almucantar with $z_0 = 149°.329254$

Table 8.1 Parameters of the Postel projection

Parameter	Options of Postel projection	
$z_0(°)$	166	149.329254
Percentage of globe surface coverage	98.5	93
$V_{b\ \min}(\%)$	– 84.5	– 67.3
$V_{a\ \min}(\%)$	– 84.5	– 67.3
$V_{b\ \max}(\%)$	– 84.5	– 67.3
$V_{a\ \max}(\%)$	84.5	67.3
$V_{p\ \min}(\%)$	– 97.6	– 89.3
$V_{p\ \max}(\%)$	– 71.5	– 45.2
$\omega_{\min}(°)$	0	0
$\omega_{\max}(°)$	115.5	84.5
Ξ_A (%)	78.3	59.0
$\Xi_{\mathrm{Ch}}(\%)$	84.5	67.3

Considering that ideal projections with a boundary almucantar $z_0 = 149.329254°$ have territorial coverage that is not much smaller than the territorial coverage of projections with a boundary almucantar of 166°, projections with a smaller almucantar can be recommended for showing the territory of the globe.

8.2 Best Conformal Projections

8.2.1 Chebyshev Projections

The great mathematician of the nineteenth century, P. Chebyshev, is considered the founder of the field of mathematics known as the theory of approximation of functions. Within this theory, he introduced the minimax criterion for measuring the proximity of one function to another. In the nineteenth century, this was a completely unusual criterion, which nevertheless became common not only in mathematics but also in cartography, particularly when solving the problem of finding the best projections.

According to the bulletin of the Imperial Academy of Sciences of St. Petersburg (Tchebychev 1856), on January 18, 1856, a meeting of the physics and mathematics class was held, at which prof. P. Chebyshev presented a report on the topic "Sur la construction des cartes géographiques". As part of this report, Chebyshev formulated the theorem for the best conformal projections, which now bears his name.

In the modern interpretation performed by G. Meshcheryakov in (1968), this theorem is stated as follows:

The best conformal projection of the ellipsoid surface onto a plane is one in which the scale factor is constant at the boundary of the region.

As shown below, P. Chebyshev proposed a conformal projection that was best according to the minimax criterion, i.e., best according to the criterion that now bears his name. He did not provide proof of the theorem in his report but promised to do so in the future.

The fact is that the best conformal projection demonstrates a special case of functions closest to zero, known as subharmonic functions. The simplest example of such a function is a quadratic function in one variable, whose graph is a parabola. Figure 8.5 shows several positions of the parabola.

Obviously, only the position in Fig. 8.5b is closest to zero within the interval of change in the value $x \in [-4, 4]$. This is exactly how the linear distortions of the Chebyshev projection behave, reaching a minimum equal to $(-H)$ at the central point inside the mapping area and a maximum equal to $(+H)$ at the boundary of the area.

Realizing the obviousness of the theorem he formulated, P. Chebyshev postponed the formalization of its proof and publication until later. As a result, the theorem has remained unproven for 38 years. It was not until 1896 that the Ukrainian cartographer and student of P. Chebyshev, Gravé, finally proved the theorem (1896, 1911).

Before presenting the proof of Chebyshev's theorem, we first analyze the basic formulae of all conformal projections.

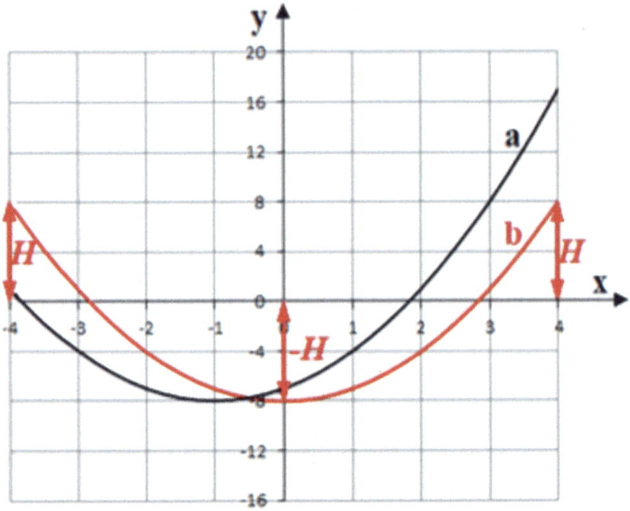

Fig. 8.5 Various positions of the parabola: graph **a** represents an arbitrary position of the parabola, while graph **b** minimizes its deviation from zero

8.2.2 *Linear Scale Factor for Conformal Projections*

As stated in Sect. 1.9.1, in conformal projections, the extremal scale factors are equal (see Formula (1.138)). Substituting this formula into (2.9–2.10) yields:

$$N_q = ra\cos\beta, \quad N_\lambda = ra\sin\beta, \tag{8.28}$$

$$E_q = -ra\sin\beta, \quad E_\lambda = ra\cos\beta. \tag{8.29}$$

Comparing Eqs. (8.28) and (8.29), we can write:

$$N_q = E_\lambda, \quad N_\lambda = -E_q. \tag{8.30}$$

Consequently, N, E are conjugate harmonic functions.

Additionally, $\ln(ra)$, $(-\beta)$ are conjugate harmonic functions:

$$\Phi'_q = -\beta_\lambda, \Phi'_\lambda = \beta_q, \tag{8.31}$$

where

$$\Phi' = \ln(ra). \tag{8.32}$$

Thus, the function Φ' satisfies Laplace's equation, namely:

$$\Delta\Phi' = \Delta \ln(ra) = \Delta \ln r + \Delta \ln a = 0. \tag{8.33}$$

Hence,

$$\Delta \ln a = -\Delta \ln r = -\Delta\tau. \tag{8.34}$$

As shown in Sect. 8.2.1, for an ellipsoid of revolution, the function $\Delta\tau \leq 0$. This means that:

$$\Delta \ln a \geq 0. \tag{8.35}$$

Functions whose Laplace operator is nonnegative are called subharmonic functions, according to (Timan and Trofimov 1968).

Let us rewrite Eq. (8.32) as follows:

$$\ln a = \Phi' - \ln r. \tag{8.36}$$

8.2.3 Chebyshev's Theorem

This theorem is proven via the Gravé method (Gravé 1911). To simplify the proof, we use the properties of harmonic and subharmonic functions, which are now considered trivial but were still the subject of deep research by scientists in 1911.

Let us outline the basic properties of harmonic and subharmonic functions (Timan and Trofimov 1968).

1. Harmonic functions reach their maximum and minimum at the boundary of the region.
2. Subharmonic functions reach a maximum at the boundary of the region.

Let the best projection according to Chebyshev's criterion for some mapping region be one that satisfies Chebyshev's theorem, i.e., at the boundary:

$$\ln a|_{\text{on borderline}} = \ln a_{\text{Ch}} = H = \text{const}. \tag{8.37}$$

Since the function $\ln a$ is subharmonic, the value of H is the maximum of the function. At some point inside the mapping area, the function $\ln a$ reaches the minimum equal to $(-H)$. The value of Chebyshev's criterion for this projection is equal to H.

Now, suppose there is a conformal projection with a logarithm of the scale factor $\ln a'$, which has a Chebyshev's criterion value less than H. This means that at the boundary:

$$\ln a'|_{\text{on borderline}} = \ln a'_0 \leq H, \tag{8.38}$$

and at some point inside the region:

$$\ln a' \geq -H. \tag{8.39}$$

Consider the harmonic function:

$$\ln a - \ln a' = (\Phi' - \ln r) - \left(\Phi'' - \ln r\right) = \Phi' - \Phi''. \tag{8.40}$$

At the boundary, this function, according to conditions (8.37–8.38), is nonnegative, namely:

$$(\ln a - \ln a')|_{\text{on borderline}} \geq 0. \tag{8.41}$$

Since the harmonic function reaches the minimum at the boundary, this minimum is nonnegative, then everywhere in the mapping area:

$$(\ln a - \ln a') \geq 0,$$

or

$$\ln a' \leq \ln a. \tag{8.42}$$

This inequality holds at all points within the mapping area, including at the point A_0 of the minimum of the function $\ln a$, namely:

$$\ln a'(A_0) \leq -H. \tag{8.43}$$

This means that the value of Chebyshev's criterion for a projection with the logarithm of the scale factor $\ln a'$ is not less than the criterion for a projection satisfying Chebyshev's theorem.

The proof of the theorem does not rely on the property of the mapping area to be simply connected or doubly connected. Therefore, the result of the theorem can be extended to a special case when the projection is performed on the surface of a cylinder or cone and the mapping area is doubly connected. However, it should be noted that the best conical conformal projection according to Chebyshev's criterion was found by A. Markov (1895). As expected, he derived a projection with two standard parallels.

8.2.4 Chebyshev's Criterion

To determine the constant H that minimizes the function *lna* in the mapping area, we use Green's formula. As mentioned above, a doubly continuously differentiable function is defined by using Eq. (3.39). Applying this equality to conformal projections satisfying Chebyshev's theorem, after identical transformations, we obtain:

$$H = \frac{1}{2} \iint G(A_0)\Delta(\ln r)d\sigma. \tag{8.44}$$

In Eq. (8.44), the Green's function $G(A_0)$ is determined by the same conditions as the Green's function defined in Sect. 3.3.6, except for the condition that it is constructed for the point A_0, where the function *lna* reaches the minimum inside the mapping area.

By analyzing Formula (8.44), we can conclude that the constant H of the Chebyshev projection depends on the Green's function constructed for the point A_0, on the Laplace operator of the coefficient of the metric shape of the surface and of the mapping area.

8.2.5 Chebyshev's Criterion for Azimuthal Stereographic Projection

Let us find the value of H for the azimuthal conformal projection, where the area is bounded by the almucantar with a zenith distance equal to z_0. The Green's function for the central point of the region, as indicated in (Timan and Trofimov 1968, p. 98), is given by:

$$G(A_0) = \frac{1}{2\pi}(\ln \rho - \ln \rho_0), \tag{8.45}$$

where ρ is the distance from the central point of the projection to an arbitrary point inside the mapping area and where ρ_0 is the distance from the central point to a point on the boundary of the mapping area, i.e., on the almucantar with a zenith distance z_0. For stereographic projection (Bugayevskiy and Snyder 1995, p. 101):

$$\rho = 2k \tan \frac{z}{2}, \tag{8.46}$$

where k is the constant of the projection.

Substituting Eq. (8.46) into Eq. (8.45) and performing identical transformations leads to:

$$G(A_0) = \frac{1}{2\pi}\left(\ln \tan \frac{z}{2} - \ln \tan \frac{z_0}{2}\right). \tag{8.47}$$

For an ellipsoid of revolution, the Laplace operator of the logarithm of the coefficient of the metric shape of the surface is represented by Formula (2.23). A similar formula for a sphere with a polar coordinate system z, α is as follows:

$$\Delta\tau = -\sin^2 z. \tag{8.48}$$

Taking into account Formula (8.48), we rewrite Eq. (8.34) for the sphere as follows:

$$\Delta \ln a = -\Delta\tau = \sin^2 z. \tag{8.49}$$

For a sphere, we can use the system of isometric coordinates q', λ', related to the polar coordinates α, z by equalities (5.2):

$$H = \frac{1}{2}\int_0^{z_0} G(A_0)\Delta(\ln r)\mathrm{d}q'\mathrm{d}\lambda'. \tag{8.50}$$

Let us replace the variable λ' with α, integrate over this variable, and move from the variable q' to the variable z. Hence,

$$H = -\frac{1}{2}\int_0^{z_0}\left(ln\tan\frac{z}{2} - ln\tan\frac{z_0}{2}\right)\sin z\mathrm{d}z. \tag{8.51}$$

Let us denote:

$$\begin{aligned} X &= \int\left(\ln\tan\frac{z}{2}\right)\sin z\mathrm{d}z \\ Y &= \int\left(\ln\tan\frac{z_0}{2}\right)\sin z\mathrm{d}z. \end{aligned} \tag{8.52}$$

Then

$$H = -\frac{1}{2}\left(X\Big|_0^{z_0} - Y\Big|_0^{z_0}\right). \tag{8.53}$$

The indefinite integral of X is equal to:

$$X = 2ln\tan\frac{z}{2}\left(-\cos^2\frac{z}{2}\right) + ln\sin^2\frac{z}{2}. \tag{8.54}$$

At $z = 0$, the function X becomes indeterminate in the form $\infty - \infty$. However, it has a limit of 0. Therefore,

$$X\Big|_0^{z_0} = \left(-2\cos^2\frac{z_0}{2}\right)\ln\tan\frac{z_0}{2} + \ln\sin^2\frac{z_0}{2}. \tag{8.55}$$

For the second integral, we have:

$$Y = -\cos z \cdot \ln\tan\frac{z_0}{2}, \tag{8.56}$$

or,

$$Y\Big|_0^{z_0} = -\cos z_0 \cdot \ln\tan\frac{z_0}{2} + \ln\tan\frac{z_0}{2}. \tag{8.57}$$

Substituting Eqs. (8.55) and (8.57) into Eq. (8.53) gives:

$$H = -\frac{1}{2}\left(\left(\cos z_0 - 2\cos^2\frac{z_0}{2} - 1\right)\ln\tan\frac{z_0}{2} + \ln\sin^2\frac{z_0}{2}\right),$$

or,

$$H = -\ln\cos\frac{z_0}{2}. \tag{8.58}$$

To verify the resulting formula, we write the equation for the linear scale factor of the azimuthal stereographic projection (Bugayevskiy and Snyder 1995, p. 103) as

$$a = \frac{k}{\cos^2\frac{z}{2}}. \tag{8.59}$$

The constant k is determined from the condition that at the boundary, $\ln a$ is equal to the constant H, and at the center, where the minimum is reached, it is equal to the constant $(-H)$, i.e.,

$$\ln a(A_0) = \ln k = -H, \tag{8.60}$$

$$\ln a_{\text{on borderline}} = \ln k - 2\ln\cos\frac{z_0}{2} = H. \tag{8.61}$$

Hence,

$$H = -\ln\cos\frac{z_0}{2}. \tag{8.62}$$

Formulae (8.58) and (8.62) completely coincide.

8.3 Markov Projection

According to (Bugayevskiy and Snyder 1995, p. 89), the formulae for direct conic projections have the following form:

$$N = -\rho \cos \delta, \quad E = \rho \sin \delta, \tag{8.63}$$

$$\rho = f(\varphi), \quad \delta = \kappa \cdot \lambda, \quad \kappa = \text{const}, \tag{8.64}$$

where ρ is the radius of the parallel with latitude φ, which is constructed on the plane of the projection. The extremal scale factors of the conical projection are equal (Bugayevskiy and Snyder 1995, p. 90):

$$m = -\frac{\rho_\varphi}{M}, \tag{8.65}$$

$$n = \kappa \frac{\rho}{r}. \tag{8.66}$$

Since all conic projections are orthogonal, the extremal scale factors satisfy conditions (1.142).

The Markov projection is the best conic projection according to Chebyshev's criterion. It is one-to-one; however, it is not twice continuously differentiable throughout the mapping region. It is a composite projection consisting of three separate projections, meaning that the entire mapping area is divided into three zones, each with its own projection. Within each zone, the corresponding projection is continuously differentiable. At the boundaries between the zones, the continuity of the function ρ and its first derivative is maintained, but its continuous differentiability of the second and subsequent derivatives is violated.

As mentioned above, the projection is constructed by using the method that is now used to approximate functions with splines. Moreover, the spline is constructed for the function ρ in such a way that it minimizes the linear distortions of the projection in the minimax sense.

Let us describe the procedure for constructing this projection.

8.3.1 Procedure for Constructing the Markov Projection

The entire mapping area is bounded by two parallels with latitudes φ_1, φ_2. By adding two more parallels with latitudes φ_3, φ_4, the mapping region can be divided into three zones, as shown in Fig. 8.6:

$$\varphi_1 < \varphi_3 < \varphi_4 < \varphi_2. \tag{8.67}$$

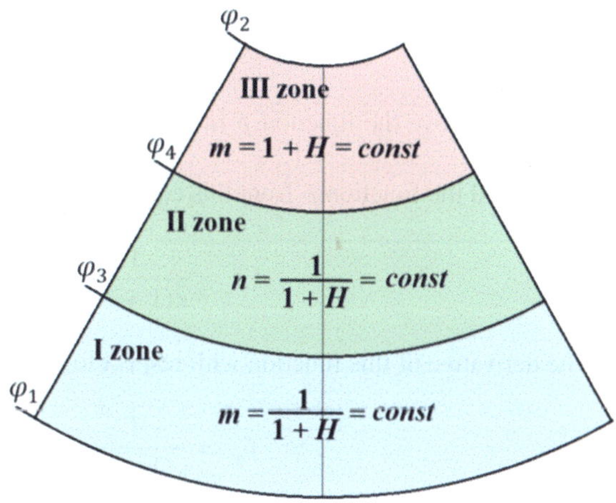

Fig. 8.6 Scheme of dividing the mapping region into three parts in the Markov projection

As shown in Fig. 8.6, in each zone, one of the extremal scale factors is equal to some constant.

First Zone

In this zone, the projection is equidistant along the meridians in the broad sense, meaning that the linear scale factor along the meridians is equal to a constant, namely:

$$m = -\frac{\dot{\rho}_\varphi}{M} = \frac{1}{1+H} = \text{const.} \tag{8.68}$$

Let us find the derivative ρ_φ from this formula:

$$\dot{\rho}_\varphi = -M\frac{1}{1+H}. \tag{8.69}$$

Integrating Formula (8.69) gives:

$$\dot{\rho} = -\frac{1}{1+H}\int_0^\varphi M\,\mathrm{d}\varphi + C_1, \tag{8.70}$$

where C_1 is the integration constant. The dot over the letter ρ in Formulae (8.68–8.70) indicates that the function ρ and its derivatives are defined for the first zone.

Second Zone

In the second zone, the Markov projection is equidistant along the parallel, also in a broad sense, meaning that the linear scale factor along the parallel is equal to a constant value, namely:

$$n = \kappa \frac{\ddot{\rho}}{r} = \frac{1}{1+H} = \text{const.} \tag{8.71}$$

Two dots above the function ρ indicate that this function is determined for the second zone.

Let us find the function $\ddot{\rho}$ from this equality.

$$\ddot{\rho} = \frac{1}{\kappa(1+H)} r. \tag{8.72}$$

The derivative of this function with respect to latitude is equal to:

$$\ddot{\rho}_{\varphi} = \frac{1}{\kappa(1+H)} r_{\varphi}. \tag{8.73}$$

Third Zone

In the third zone, as well as in the first zone, the projection is equidistant along the meridians in a broad sense. However, the linear scale factor along the meridians is equal to a different constant:

$$m = -\frac{\dddot{\rho}_{\varphi}}{M} = (1+H) = \text{const.} \tag{8.74}$$

For the derivative of the function $\dddot{\rho}_{\varphi}$, we have:

$$\dddot{\rho}_{\varphi} = -(1+H)M. \tag{8.75}$$

Integrating Formula (8.75) gives:

$$\dddot{\rho} = -(1+H)\int_{0}^{\varphi} M \, \mathrm{d}\varphi + C_2, \tag{8.76}$$

where C_2 is the integration constant. Three dots above the letter ρ indicate that this value is determined for points of the third zone.

8.3.2 Projection Parameters

The projection has six parameters:

$$C_1, C_2, \varphi_3, \varphi_4, \kappa, (1+H).$$

There are six conditions to determine them. Let us write them down using definitions adopted in spline theory.

Boundary Conditions

At the boundary of the mapping area, on parallels with latitudes φ_1, φ_2, the linear scale factors along the parallels are equal to:

$$n_1 = n_2 = 1 + H, \tag{8.77}$$

or,

$$\kappa \frac{\dot{\rho}(\varphi_1)}{r(\varphi_1)} = \kappa \frac{\dddot{\rho}(\varphi_2)}{r(\varphi_2)} = 1 + H. \tag{8.78}$$

In equalities (8.78), $\dot{\rho}(\varphi_1), r(\varphi_1)$ are the values of the functions $\dot{\rho}, r$ on the boundary parallel with latitude φ_1. Similarly, $\dddot{\rho}(\varphi_2), r(\varphi_2)$ are the values of the functions $\dddot{\rho}, r$ on the boundary parallel with latitude φ_2.

Let us find the quantities $\dot{\rho}(\varphi_1), \dddot{\rho}(\varphi_2)$ from equalities (8.78).

$$\dot{\rho}(\varphi_1) = \frac{1}{\kappa}(1 + H) r(\varphi_1), \tag{8.79}$$

$$\dddot{\rho}(\varphi_2) = \frac{1}{\kappa}(1 + H) r(\varphi_2). \tag{8.80}$$

Docking Conditions

The docking conditions are divided into two groups:

Conditions for joining the function ρ between neighboring zones along parallels with latitudes φ_3, φ_4:

$$\dot{\rho}(\varphi_3) = \ddot{\rho}(\varphi_3), \ddot{\rho}(\varphi_4) = \dddot{\rho}(\varphi_4). \tag{8.81}$$

Conditions for joining the derivatives of the function ρ between neighboring zones along parallels with latitudes φ_3, φ_4.

$$\dot{\rho}_\varphi(\varphi_3) = \ddot{\rho}_\varphi(\varphi_3), \ddot{\rho}_\varphi(\varphi_4) = \dddot{\rho}_\varphi(\varphi_4). \tag{8.82}$$

Using boundary conditions (8.78) or the conditions in (8.81) for connecting the function ρ between neighboring zones along parallels with latitudes φ_3, φ_4 allows one to find the integration constants. Since the obtained equalities will be needed later, we will determine the constants C_1, C_2 using two different formulae.

Determination of the Constant C_1

Using Formula (8.70), we find the value of the function $\dot{\rho}$ on the parallel with latitude φ_1.

$$\dot{\rho}(\varphi_1) = -\frac{1}{1+H}\int\limits_0^{\varphi_1} M\,\mathrm{d}\varphi + C_1. \tag{8.83}$$

Taking into account Eq. (8.79), the constant C_1 is equal to:

$$C_1 = \frac{1}{1+H}\int\limits_0^{\varphi_1} M\,\mathrm{d}\varphi + \frac{(1+H)}{\kappa}r(\varphi_1). \tag{8.84}$$

Similarly, using Formula (8.70), we find the value of the function $\dot{\rho}$ on a parallel with latitude φ_3.

$$\dot{\rho}(\varphi_3) = -\frac{1}{1+H}\int\limits_0^{\varphi_3} M\,\mathrm{d}\varphi + C_1. \tag{8.85}$$

Using the first formula of the system of equalities (8.81) and Formulae (8.72 and 8.85), we write the following equation for the constant C_1.

$$C_1 = \frac{1}{1+H}\int\limits_0^{\varphi_3} M\,\mathrm{d}\varphi + \frac{1}{\kappa(1+H)}r(\varphi_3). \tag{8.86}$$

Combining equalities (8.84) and (8.86), we obtain:

$$\alpha\int\limits_{\varphi_1}^{\varphi_3} M\,\mathrm{d}\varphi + r(\varphi_3) - (1+H)^2 r(\varphi_1) = 0. \tag{8.87}$$

Thus, the function ρ for the first zone can be determined by one of the following formulae:

$$\dot{\rho} = -\frac{1}{1+H}\int\limits_{\varphi_1}^{\varphi} M\,\mathrm{d}\varphi + \frac{(1+H)}{\kappa}r(\varphi_1). \tag{8.88}$$

$$\dot{\rho} = \frac{1}{1+H}\int\limits_{\varphi}^{\varphi_3} M\,\mathrm{d}\varphi + \frac{1}{\kappa(1+H)}r(\varphi_3). \tag{8.89}$$

Determination of the Constant C_2

Similarly, using the Formulae (8.76 and 8.80) for the parallel with latitude φ_2, we obtain the first expression for the constant C_2.

$$C_2 = (1+H)\int\limits_0^{\varphi_2} M\,\mathrm{d}\varphi + \frac{1}{\kappa}(1+H)r(\varphi_2). \tag{8.90}$$

The second condition of the system (8.81), taking into account (8.72 and 8.76), allows us to write:

$$C_2 = (1+H)\int\limits_0^{\varphi_4} M\,\mathrm{d}\varphi + \frac{1}{\kappa(1+H)}r(\varphi_4). \tag{8.91}$$

Combining equalities (8.90–8.91), we obtain:

$$\kappa\int\limits_{\varphi_4}^{\varphi_2} M\,\mathrm{d}\varphi + r(\varphi_2) - \frac{1}{(1+H)^2}r(\varphi_4) = 0. \tag{8.92}$$

Considering Eqs. (8.76, 8.90–8.91), we write the following formulae to determine ρ for the third zone:

$$\dddot{\rho} = (1+H)\int\limits_{\varphi}^{\varphi_2} M\,\mathrm{d}\varphi + \frac{1}{\kappa}(1+H)r(\varphi_2), \tag{8.93}$$

$$\dddot{\rho} = -(1+H)\int\limits_{\varphi_4}^{\varphi} M\,\mathrm{d}\varphi + \frac{1}{\kappa(1+H)}r(\varphi_1). \tag{8.94}$$

Conditions for Joining Derivatives of the Function ρ

Substituting (8.69 and 8.73) into the first equality of system (8.82) yields the condition for joining the derivatives of the function ρ between zones I and II along the parallel with latitude φ_3:

$$\kappa = -\frac{1}{M_3}r_\varphi(\varphi_3). \tag{8.95}$$

Taking into account the formulae for M, r_φ (see Eq. (1.8) and Appendix A) for latitude φ_3, we get:

$$\kappa = \sin\varphi_3. \tag{8.96}$$

Similarly, the condition for joining the derivatives of the function ρ between zones II and III along the parallel with latitude φ_4, is as follows:

$$\kappa(1+H)^2 = -\frac{1}{M_4} r_\varphi(\varphi_4). \tag{8.97}$$

The substitution of (8.95) into Eq. (8.97) gives:

$$(1+H)^2 = \frac{M_3}{M_4} \cdot \frac{r_\varphi(\varphi_4)}{r_\varphi(\varphi_3)}, \tag{8.98}$$

or taking into account Eq. (8.96):

$$(1+H)^2 = \frac{\sin\varphi_4}{\sin\varphi_3}. \tag{8.99}$$

Substituting the equalities for κ and $(1+H)^2$ from Eqs. (8.95 and 8.98) into Eqs. (8.87 and 8.92) leads to:

$$\frac{1}{M_3} r_\varphi(\varphi_3) \int\limits_{\varphi_1}^{\varphi_3} M\,\mathrm{d}\varphi + \frac{M_3}{M_4} \cdot \frac{r_\varphi(\varphi_4)}{r_\varphi(\varphi_3)} r_1 - r_3 = 0, \tag{8.100}$$

$$-\frac{1}{M_3} r_\varphi(\varphi_3) \int\limits_{\varphi_4}^{\varphi_2} M\,\mathrm{d}\varphi + \frac{M_4}{M_3} \cdot \frac{r_\varphi(\varphi_3)}{r_\varphi(\varphi_4)} r_2 - r_4 = 0. \tag{8.101}$$

Equations (8.100–8.101) form a system of two transcendental equations with two unknowns φ_3, φ_4, which can be solved by one of the known approximate methods. In particular, the use of Newton's method made it possible to find a projection, which will be discussed below in Sect. 8.3.5.

8.3.3 Linear Scale Factors Along the Parallels with Latitudes φ_3, φ_4

Formulae (8.100–8.101) can also be obtained from the condition that, along the parallels with latitudes φ_3, φ_4, the function n has a derivative equal to zero on both the left and right sides of zone II. Indeed, since ρ is not only a continuous function but also has a continuous first derivative, the scale factor n is also continuous everywhere in the mapping area, along with its first derivative. In the latitudinal interval from φ_3 to φ_4, the linear scale factor is equal to the constant $\frac{1}{1+H}$ (see Eq. (8.71) and Fig. 8.7). This means that at each point of this interval, the derivative of the function n is equal to zero.

In the interval $\varphi_1 \le \varphi \le \varphi_3$ (1 zone), the linear scale factor along the parallels, taking into account Formulae (8.66 and 8.88), is equal to:

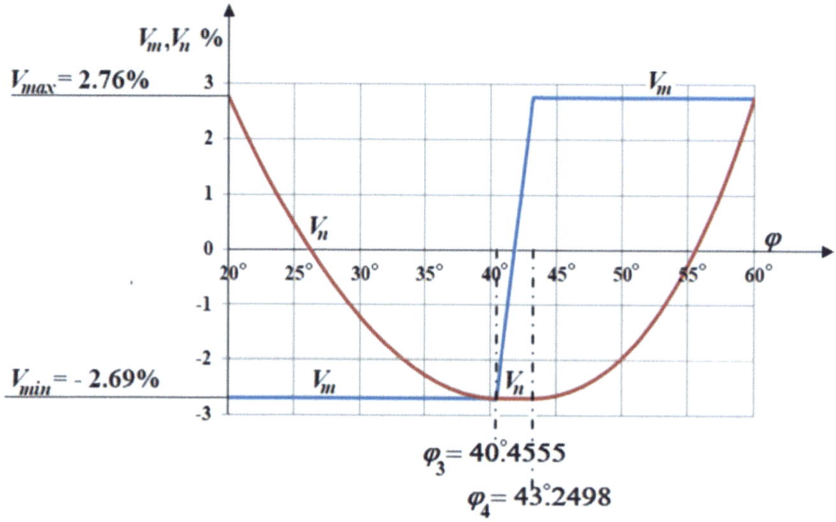

Fig. 8.7 Graphs of changes in linear distortions along the meridians (blue) and parallels (red) of the Markov projection, calculated using the formula $V = \mu - 1$

$$n = \alpha \frac{\dot{\rho}}{r} = \frac{1}{1+H} \frac{-\kappa \int_{\varphi_1}^{\varphi} M \, \mathrm{d}\varphi + (1+H)^2 r(\varphi_1)}{r}. \tag{8.102}$$

Differentiating this formula gives:

$$\frac{\mathrm{d}n}{\mathrm{d}\varphi} = \frac{1}{1+H} \frac{(-\kappa M r - \left(-\kappa \int_{\varphi_1}^{\varphi} M d\varphi + (1+H)^2 r(\varphi_1)\right) r_\varphi \Big)}{r^2}. \tag{8.103}$$

In parallel with latitude φ_3, Eq. (8.103) will look like this:

$$\frac{\mathrm{d}n}{\mathrm{d}\varphi}(\varphi_3) = \frac{1}{(1+H)} \frac{(-\kappa M_3 r(\varphi_3) + \kappa r_\varphi(\varphi_3) \int_{\varphi_1}^{\varphi_3} M \, \mathrm{d}\varphi - (1+H)^2 r(\varphi_1) r_\varphi(\varphi_3))}{r^2}. \tag{8.104}$$

Let us divide the numerator and denominator of Eq. (8.104) by κM_3. Taking into account (8.95 and 8.98), we write:

$$\frac{\mathrm{d}n}{\mathrm{d}\varphi}(\varphi_3) = \frac{\kappa M_3}{(1+H)} \frac{\left[-r(\varphi_3) + \frac{r_\varphi(\varphi_3)}{M_3} \int_{\varphi_1}^{\varphi_3} M \, \mathrm{d}\varphi + \frac{M_3}{M_4} \cdot \frac{r_\varphi(\varphi_4)}{r_\varphi(\varphi_3)} r(\varphi_1)\right]}{r^2}. \tag{8.105}$$

The comparison of the expression in the numerator in square brackets with Eq. (8.100) allows us to conclude the following:

$$\frac{dn}{d\varphi}(\varphi_3) = 0. \tag{8.106}$$

The value of the derivative of the linear scale factor along the parallel with latitude φ_4 is determined similarly.

This means that Eqs. (8.100–8.101) reflect the property of the linear scale factor along a parallel to have a minimum on parallels with latitudes φ_3, φ_4.

8.3.4 Markov's Theorem

Using modern definitions, **Markov's theorem** on the best conic projections can be formulated as follows:

For the region of the ellipsoid of revolution bounded by two parallels with latitudes φ_1, φ_2, all conic projections have a linear scale factor no less than $(1 + H)$ and no more than $\frac{1}{1+H}$, where H is determined by formula (8.99), taking into account (8.100–8.101).

A. Markov (1895) noted that the projection he proposed is the best conical projection for a surface of revolution, for which the second derivative of the metric coefficient with respect to latitude is negative in the mapping area. These surfaces include an ellipsoid of revolution and a sphere (see equalities (2.25 and 8.48)). However, the proof of the theorem proposed by A. Markov is rigorous only for the sphere. Let us consider this proof in detail.

According to (Markov 1895), if there was a projection with less linear distortion, then it would satisfy the following inequalities:

1. For the first zone with $\varphi_1 \leq \varphi \leq \varphi_3$:

$$m = -\frac{\rho_\varphi}{M} > \frac{1}{1+H}. \tag{8.107}$$

From Eq. (8.107), we obtain:

$$-\rho_\varphi > M\frac{1}{1+H}. \tag{8.108}$$

2. For the third zone with $\varphi_4 \leq \varphi \leq \varphi_2$:

$$m = -\frac{\rho_\varphi}{M} < 1 + H, \tag{8.109}$$

or

$$-\rho_\varphi < M(1+H). \tag{8.110}$$

Moreover, in both intervals, the linear scale factor along the parallels must satisfy the following inequalities:

$$n = \overline{\kappa}\frac{\rho}{r}\Big\langle 1 + H, n = \overline{\kappa}\frac{\rho}{r}\Big\rangle\frac{1}{1+H}. \tag{8.111}$$

In Eq. (8.111), $\overline{\kappa}$ is the parameter of the projection in question. A priori:

$$\overline{\kappa} \neq \kappa,$$

where κ is the parameter of the best conic projection, defined by Formulae (8.68–8.101).

From the last inequalities for the interval $\varphi_1 \leq \varphi \leq \varphi_3$, we have:

$$\rho(\varphi_1) < \frac{1}{\overline{\kappa}}(1 + H)r(\varphi_1), \tag{8.112}$$

$$\rho(\varphi_3) > \frac{1}{\overline{\kappa}}\frac{1}{(1 + H)}r(\varphi_3). \tag{8.113}$$

Taking into account inequalities (8.112–8.113), we can write:

$$\rho(\varphi_1) - \rho(\varphi_3) < \frac{1}{\overline{\kappa}}(1 + H)r(\varphi_1) - \frac{1}{\overline{\kappa}}\frac{1}{(1 + H)}r(\varphi_3). \tag{8.114}$$

Let us turn to inequality (8.108), according to which the derivative of the function ρ, taken with a minus sign, at all points of the interval $\varphi_1 \leq \varphi \leq \varphi_3$ takes on a value greater than $M\,\frac{1}{1+H}$. This means that the finite differences of this function are also greater than the value of $M\,\frac{1}{1+H}$, i.e.:

$$-\frac{\rho(\varphi_3) - \rho(\varphi_1)}{\varphi_3 - \varphi_1} > M\,(\varphi_5)\frac{1}{1 + H}, \tag{8.115}$$

or,

$$\rho(\varphi_1) - \rho(\varphi_3) > \frac{1}{1 + H}M\,(\varphi_5)(\varphi_3 - \varphi_1). \tag{8.116}$$

In Formula (8.116), the function $M\,(\varphi_5)$ takes on a certain value for the latitude interval $\varphi_1 \leq \varphi \leq \varphi_3$, that is, $\varphi_1 \leq \varphi_5 \leq \varphi_3$.

Comparing inequalities (8.114) and (8.116), we obtain:

$$\frac{1}{1 + H}M\,(\varphi_5)(\varphi_3 - \varphi_1) < \frac{1}{\overline{\kappa}}(1 + H)r(\varphi_1) - \frac{1}{\overline{\kappa}}\frac{1}{(1 + H)}r(\varphi_3), \tag{8.117}$$

or, after identical transformations:

$$\overline{\kappa} < \frac{1}{M(\varphi_5)(\varphi_3 - \varphi_1)}[(1+H)^2 r(\varphi_1) - r(\varphi_3)]. \tag{8.118}$$

According to Eq. (8.87), the expression in square brackets of Formula (8.118) is equal to:

$$(1+H)^2 r(\varphi_1) - r(\varphi_3) = \kappa \int\limits_{\varphi_1}^{\varphi_3} M \, d\varphi. \tag{8.119}$$

Thus,

$$\overline{\kappa} < \kappa \frac{\int_{\varphi_1}^{\varphi_3} M \, d\varphi}{M(\varphi_5)(\varphi_3 - \varphi_1)}. \tag{8.120}$$

Similarly, for the interval $\varphi_4 \leq \varphi \leq \varphi_2$, inequalities (8.111) can be written as follows:

$$\rho(\varphi_2) < \frac{1}{\overline{\kappa}}(1+H) r(\varphi_2), \tag{8.121}$$

$$\rho(\varphi_4) > \frac{1}{\overline{\kappa}} \frac{1}{(1+H)} r(\varphi_4). \tag{8.122}$$

Taking the difference between the values $\rho(\varphi_4) - \rho(\varphi_2)$, we obtain:

$$\rho(\varphi_4) - \rho(\varphi_2) > \frac{1}{\overline{\kappa}} \frac{1}{(1+H)} r(\varphi_4) - \frac{1}{\overline{\kappa}}(1+H) r(\varphi_2). \tag{8.123}$$

Considering that in this interval, the derivative of the function ρ satisfies inequality (8.110), we can write the following condition for finite differences:

$$-\frac{\rho(\varphi_4) - \rho(\varphi_2)}{\varphi_4 - \varphi_2} < M(\varphi_6)(1+H), \tag{8.124}$$

or,

$$\rho(\varphi_4) - \rho(\varphi_2) < M(\varphi_6)(1+H)(\varphi_2 - \varphi_4), \tag{8.125}$$

where $\varphi_4 \leq \varphi_6 \leq \varphi_2$.

Combining inequalities (8.123) and (8.125), we obtain:

$$M(\varphi_6)(1+H)(\varphi_2 - \varphi_4) > \frac{1}{\overline{\kappa}} \frac{1}{(1+H)} r(\varphi_4) - \frac{1}{\overline{\kappa}}(1+H) r(\varphi_2), \tag{8.126}$$

or,

$$\overline{\kappa} > \frac{1}{M(\varphi_6)(\varphi_2 - \varphi_4)}\left[\frac{1}{(1+H)^2}r(\varphi_4) - r(\varphi_2)\right]. \tag{8.127}$$

The expression in square brackets, according to Eq. (8.92), is equal to:

$$\frac{1}{(1+H)^2}r(\varphi_4) - r(\varphi_2) = \kappa\int\limits_{\varphi_4}^{\varphi_2} M\,\mathrm{d}\varphi. \tag{8.128}$$

Substituting Eq. (8.125) into (8.127) yields:

$$\overline{\kappa} > \kappa\frac{\int_{\varphi_4}^{\varphi_2} M\,\mathrm{d}\varphi}{M(\varphi_6)(\varphi_2 - \varphi_4)}. \tag{8.129}$$

Combining inequalities (8.120) and (8.129) gives:

$$\kappa\frac{\int_{\varphi_4}^{\varphi_2} M\,\mathrm{d}\varphi}{M(\varphi_6)(\varphi_2 - \varphi_4)} < \overline{\kappa} < \kappa\frac{\int_{\varphi_1}^{\varphi_3} M\,\mathrm{d}\varphi}{M(\varphi_5)(\varphi_3 - \varphi_1)}. \tag{8.130}$$

Applying the mean value theorem (Bronstejn et al. 2020) to the definite integrals in expression (8.130), we obtain:

$$\kappa\frac{M(\varphi_8)(\varphi_2 - \varphi_4)}{M(\varphi_6)(\varphi_2 - \varphi_4)} < \overline{\kappa} < \kappa\frac{M(\varphi_7)(\varphi_3 - \varphi_1)}{M(\varphi_5)(\varphi_3 - \varphi_1)}. \tag{8.131}$$

On the right side of inequality (8.131), the value $M(\varphi_7)$ is the value of the integrand of the integral function at some point in the interval $\varphi_1 \le \varphi \le \varphi_3$. Similarly, $M(\varphi_8)$ is the integrand on the left side of the inequality at some point in the interval $\varphi_4 \le \varphi \le \varphi_2$.

The latitudes φ_6 and φ_8 are in the same latitude interval, namely, $\varphi_4 \le \varphi \le \varphi_2$. However, it is unknown which of these latitudes is larger and which is smaller. Therefore, it is impossible to estimate the ratio $\frac{M(\varphi_8)}{M(\varphi_6)}$. A similar situation occurs with the ratio $\frac{M(\varphi_7)}{M(\varphi_5)}$.

Since (see Formula (1.8)):

$$M_{\min}|_{\varphi=0} = a\left(1 - e^2\right), \quad M_{\max}|_{\varphi=90^\circ} = \frac{a}{\sqrt{1 - e^2}},$$

then

$$0.99 \le \frac{M(\varphi_8)}{M(\varphi_6)} \le 1.01, \quad 0.99 \le \frac{M(\varphi_7)}{M(\varphi_5)} \le 1.01.$$

As a result, there is uncertainty in inequality (8.131), which does not allow for a rigorous proof of Markov's theorem for an ellipsoid of revolution.

For the sphere, the inequality (8.131) is written as follows:

$$\kappa < \overline{\kappa} < \kappa. \tag{8.132}$$

The contradiction in inequality (8.132) indicates that there is no conic projection for the surface of a sphere with linear distortions less than those of the Markov projection.

8.3.5 *Markov Projection*

The initial data for constructing the projection are the boundary parallels of the mapping area. For the resulting projection to be compared with already used conic projections, a projection with boundary parallels whose latitudes are equal:

$$\varphi_1 = 20^\circ, \quad \varphi_2 = 60^\circ, \tag{8.133}$$

was chosen.

In particular, the Lambert conformal conic projection with these boundary parallels is used to display the territory of Asia and North America. Notably, the Albers equal-area conic projection for the same areas is also used with the boundary parallels defined by Eq. (8.133). All calculations are performed for an ellipsoid with a semi-major axis equal to 1 and an eccentricity equal to that of the WGS84 ellipsoid.

The working formulae for constructing the Markov projection are given in Appendix H. Table 8.2 presents the numerical values of the quantities φ_3, φ_4, as well as κ, $(1 + H)$ for the Markov projection limited by parallels with latitudes as defined by Eq. (8.133).

Figure 8.7 shows graphs of linear distortions as a function of latitude in the Markov projection.

Linear distortions were calculated using the first formula of equalities (2.24). Since A. Markov uses a rather unique criterion to determine his projection (see Sect. 2.2), the maximum linear distortion in his projection is not equal in absolute value to the minimum distortion. The maximum linear distortion is plus 2.76%, whereas the minimum is minus 2.69%.

Table 8.2 Basic parameters of the Markov projection

Parameter	Value	Parameter	Value
$\varphi_1(^\circ)$	20	$\varphi_3(^\circ)$	40.4555271720072
$\varphi_2(^\circ)$	60	$\varphi_4(^\circ)$	43.2498233197404
κ	0.6488576277097	$1 + \mathrm{H}$	1.0276089249310

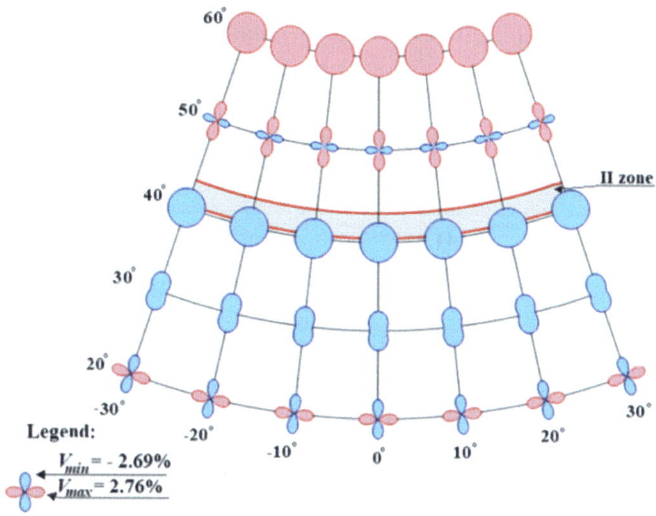

Fig. 8.8 Map grid of the Markov projection with symbol-charts of the linear distortions

Figure 8.8 shows the Markov projection map grid. Since the linear distortions in the projection are small, it is impossible to clearly display them using Tissot indicatrixes. Therefore, linear distortion symbol-charts were used. Zone II is highlighted with boundary parallels in red and shaded in gray.

Since Zone II is central and connects the first and third zones and since, according to Fig. 8.7, the distortions in this zone change very rapidly, Fig. 8.9 shows a fragment of the cartographic grid and symbol-charts of distortions of the Markov projection for this zone only. As the boundary parallels of Zone II are not multiples of integer degrees, parallels were selected where the difference in longitude is the same and equal to:

$$\Delta\varphi = \frac{\varphi_4 - \varphi_3}{4}. \tag{8.134}$$

In Fig. 8.9, the latitudes of these parallels are shown with an accuracy of 0.0001°.

To compare the linear distortions of the Markov projection with those of the most common conic projections, distortion calculations were also performed for the Lambert, Albers, and equidistant conic projections.

The above projections were determined using the minimax criterion proposed by A. Markov. This means that the extremal scale factors must satisfy the following relationships:

$$a_{\max} = 1 + V_M, \quad b_{\min} = \frac{1}{1 + V_M}. \tag{8.135}$$

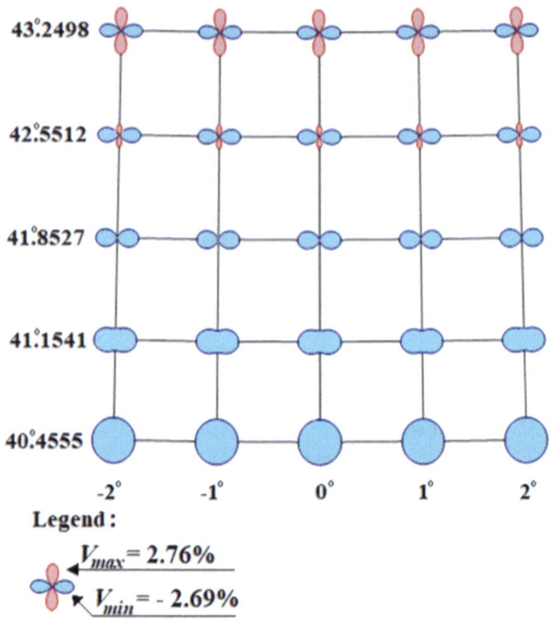

Fig. 8.9 Fragment of the Markov projection map grid for Zone II with symbol-charts of linear distortions

Appendix I provides the working formulae for calculating the parameters of the Lambert, Albers, and equidistant conic projections that satisfy these conditions.

Table 8.3 shows the extremal distortions of the Markov projection and the abovementioned conic projections.

Table 8.3 shows that the Markov projection is indeed the best conic projection according to the Chebyshev–Markov criterion (see Eq. (2.37)). The maximum linear

Table 8.3 Parameters and distortions of the Markov projection and the best projections according to the Chebyshev–Markov criterion: conformal, equal-area, equidistant

Parameter	Markov projection	Lambert projection	Albers projection	Equidistant along meridians projection
κ	0.648857627710	0.692014878853	0.645981095930	0.653744858867
k	–	1.755599390073	1.068920275311	1.832782449495
$\varphi_0(°)$	–	43.78981637	40.23927458	42.13077732
$V_{a\ \max}\%$	2.76	2.91	3.05	3.15
$V_{b\ \min}\%$	2.69	– 2.83	– 2.96	– 3.05
$V_{p\ \max}\%$	5.60	5.82	0	3.15
$V_{p\ \min}\%$	– 5.30	– 5.65	0	– 3.05
$\omega_{\max}(°)$	3.12	0	3.44	1.78
$\omega_{\min}$	0	0	0	0
$V_M\%$	2.76	2.91	3.05	3.15

distortion in this projection is 0.15% less than that in the best conformal projection, 0.29% less than that in the best equal-area projection, and 0.39% less than that in the best equidistant projection. However, this reduction in distortion is achieved by abandoning the continuous differentiability of the mapping functions. Perhaps this is why this projection has not been widely applied. Nevertheless, from a historical perspective, this may be an example of the first use of splines to approximate functions and, in fact, the first use of splines to construct map projections.

References

Bronstejn IN, Semendyaev KA, Nickel H, Marsolek L, Brunner J et al (2020) *Taschenbuch der Mathematik.* 11, aktualisierte Auflage. Haan-Gruiten, p 1280. ISBN 978-3-8085-5792-1

Bugayevskiy LM, Snyder JP (1995) Map projections. Reference Manual. Taylor&Francis Ltd., London, p 328. https://doi.org/10.1201/b16431

Gravé DA (1896) Sur la construction des Cartes geographiques. Journal de mathematiques pures et appliquees, 5(2): 317–361. http://www.numdam.org/item?id=JMPA_1896_5_2__317_0

Gravé D (1911) Demonstration d'un theoreme de Tchebychef generalize. Journal fur die reine und angewandte Mathematik. Band 140. Heft 4, 34:247–251. http://resolver.sub.uni-goettingen.de/purl?PPN243919689_0140

Markov AA (1895) O naivygodneyshikh izobrazheniyakh nekotoroy chasti dannoy poverkhnosti vrashcheniya na ploskosti (On the most advantageous images of a certain part of a given surface of rotation on a plane), Izvestiya Imperatorskoy Akademii Nauk 2(3):177–187. https://www.mathnet.ru/rus/im4207

Meshcheryakov GA (1968) Teoreticheskiye osnovy mathematicheskoy kartografii (Theoretical Bases of Mathematical Cartography). Nedra, Moskow, p 160

Novikova EN (1989) Otyskaniye ideal'nykh po kriteriyu Chebysheva proyektsiy dlya oblastey proizvol'nykh ochertaniy (Finding ideal projections according to the Chebyshev criterion for areas of arbitrary shape). Krivoy Rog, p 34. Bibliogr. 6 nazv. Rus. Dep. v UkrNIINTI, 16.10.1989, N 2214 Uk 89Dep

Tchebychev P (1856) Sur la construction des cartes géographiques. Notes 13. (Lu le 18 janvier 1856). Bulletin de la Classe Phisico-Mathématique de L'Académie Impériale des Sciences de Saint-Pétersbourg T. XIV 17(529):257–261. https://www.biodiversitylibrary.org/item/173056#page/629/mode/1up

Timan AF, Trofimov VN (1968) Vvedeniye v teoriyu garmonicheskikh funktsiy (Introduction to the Theory of Harmonic Functions). Nedra, Moskow, p 208

Appendixes

Appendix A: Radius of a Parallel, Its Logarithm, and Derivatives of These Functions

A.1 Radius of Parallel and Its Derivatives

The radius of the parallel of the ellipsoid of revolution is given by (Bagratuni 1962):

$$r = \frac{a\sqrt{1 + e'^2}\cos\varphi}{\sqrt{1 + e'^2\cos^2\varphi}}, \tag{A.1}$$

where a is the semi-major axis of the ellipsoid.

The following notations are used for the derivatives of the function r:

$$\frac{\mathrm{d}r}{\mathrm{d}q} = r_q, \quad \frac{\mathrm{d}^2r}{\mathrm{d}q^2} = r_{qq}, \quad \frac{\mathrm{d}^ir}{\mathrm{d}q^i} = r_{qi}, \quad (i = 3, 4\ldots), \tag{A.2}$$

as well as notations for the following functions of geodetic latitude:

$$\tilde{S} = \sin\varphi, \ \tilde{C} = \cos\varphi. \tag{A.3}$$

Sequentially differentiating the Eq. (A.1) under the condition (Bugayevskiy and Snyder 1995, p. 12):

$$\varphi_q = \cos\varphi\left(1 + \eta^2\right), \tag{A.4}$$

gives:

$$r_q = -r\sin\varphi, \tag{A.5}$$

E. Novikova, *Best Map Projections*, https://doi.org/10.1007/978-3-031-78334-0

$$r_{qq} = r\left(\sin^2\varphi - \cos^2\varphi - \eta^2\cos^2\varphi\right), \tag{A.6}$$

$$r_{q3} = r\sin\varphi\left(-\sin^2\varphi + 5\cos^2\varphi + 9\eta^2\cos^2\varphi + 4\eta^4\cos^2\varphi\right), \tag{A.7}$$

$$\begin{aligned} r_{q4} = r\big(&5\tilde{C}^4 - 18\tilde{S}^2\tilde{C}^2 + \tilde{S}^4 + \eta^2\tilde{C}^2\left(14\tilde{C}^2 - 58\tilde{S}^2\right) \\ &+ \eta^4\tilde{C}^2\left(13\tilde{C}^2 - 64\tilde{S}^2\right) + \eta^6\tilde{C}^2\left(4\tilde{C}^2 - 24\tilde{S}^2\right)\big), \end{aligned} \tag{A.8}$$

$$\begin{aligned} r_{q5} = r\tilde{S}\big(&-61\tilde{C}^4 + 58\tilde{S}^2\tilde{C}^2 - \tilde{S}^4 + e'^2\tilde{C}^4\left(-270\tilde{C}^2 + 330\tilde{S}^2\right) \\ &+ e'^4\tilde{C}^6\left(-445\tilde{C}^2 + 680\tilde{S}^2\right) + e'^6\tilde{C}^8\left(-324\tilde{C}^2 + 600\tilde{S}^2\right) \\ &+ e'^8\tilde{C}^{10}\left(-88\tilde{C}^2 + 192\tilde{S}^2\right)\big), \end{aligned} \tag{A.9}$$

$$\begin{aligned} r_{q6} = r\big(&-61\tilde{C}^6 + 479\tilde{S}^2\tilde{C}^4 - 179\tilde{S}^4\tilde{C}^2 + \tilde{S}^6 \\ &+ e'^2\tilde{C}^4\left(-331\tilde{C}^4 + 3298\tilde{C}^2\tilde{S}^2 - 1771\tilde{S}^4\right) \\ &+ e'^4\tilde{C}^6\left(-715\tilde{C}^4 + 8655\tilde{S}^2\tilde{C}^2 - 6080\tilde{S}^4\right) \\ &+ e'^6\tilde{C}^8\left(-769\tilde{C}^4 + 10,964\tilde{S}^2\tilde{C}^2 - 9480\tilde{S}^4\right) \\ &+ e'^8\tilde{C}^{10}\left(-412\tilde{C}^4 + 6760\tilde{S}^2\tilde{C}^2 - 6912\tilde{S}^2\right) \\ &+ e'^{10}\tilde{C}^{12}\left(-88\tilde{C}^4 + 1632\tilde{S}^2\tilde{C}^2 - 1920\tilde{S}^4\right)\big) \end{aligned} \tag{A.10}$$

$$\begin{aligned} r_{q7} = r\tilde{S}\big(&1385\tilde{C}^6 - 3111\tilde{S}^2\tilde{C}^4 + 543\tilde{S}^4\tilde{C}^2 - \tilde{S}^6 \\ &+ e'^2\tilde{C}^4\left(10,899\tilde{C}^4 - 32,802\tilde{S}^2\tilde{C}^2 + 9219\tilde{S}^4\right) \\ &+ e'^4\tilde{C}^6\left(34,419\tilde{C}^4 - 129,087\tilde{S}^2\tilde{C}^2 + 49,644\tilde{S}^4\right) \\ &+ e'^6\tilde{C}^8\left(56,385\tilde{C}^4 - 252,084\tilde{S}^2\tilde{C}^2 + 121,800\tilde{S}^4\right) \\ &+ e'^8\tilde{C}^{10}\left(50,856\tilde{C}^4 - 263,088\tilde{S}^2\tilde{C}^2 + 151,872\tilde{S}^4\right) \\ &+ e'^{10}\tilde{C}^{12}\left(24,048\tilde{C}^4 - 140,928\tilde{S}^2\tilde{C}^2 + 94,080\tilde{S}^4\right) \\ &+ e'^{12}\tilde{C}^{14}\left(4672\tilde{C}^4 - 30,528\tilde{S}^2\tilde{C}^2 + 23,040\tilde{S}^4\right)\big). \end{aligned} \tag{A.11}$$

A.2 Derivatives of the Logarithm of the Radius of a Parallel

Differentiation of Eq. (1.65) gives:

$$\tau_q = \frac{r_q}{r}. \tag{A.12}$$

Taking into account Eq. (A.5):

$$\tau_q = -\sin\varphi. \tag{A.13}$$

Successive differentiation of Formula (A.13) gives:

$$\tau_{qq} = -\left(1+\eta^2\right)\cos^2\varphi, \tag{A.14}$$

$$\tau_{q3} = 2\left(1+\eta^2\right)\left(1+2\eta^2\right)\sin\varphi\cos^2\varphi, \tag{A.15}$$

$$\tau_{q4} = 2\tilde{C}^2\left(1+\eta^2\right)\left(\tilde{C}^2-2\tilde{S}^2+3\eta^2\left(\tilde{C}^2-4\tilde{S}^2\right)+2\eta^4\left(\tilde{C}^2-6\tilde{S}^2\right)\right), \tag{A.16}$$

$$\tau_{q5} \cong -8\tilde{S}\tilde{C}^2\left(1+\eta^2\right)\left(2\tilde{C}^2-\tilde{S}^2+\eta^2\left(13\tilde{C}^2-14\tilde{S}^2\right)+\eta^4\left(22\tilde{C}^2-36\tilde{S}^2\right)\right), \tag{A.17}$$

$$\begin{aligned}\tau_{q6} \cong &-8\tilde{C}^2\left(1+\eta^2\right)\left(2\tilde{C}^4-11\tilde{S}^2\tilde{C}^2+2\tilde{S}^4+15\eta^2\left(\tilde{C}^4-9\tilde{S}^2\tilde{C}^2+4\tilde{S}^4\right)\right.\\ &\left.+\,5\eta^4\left(7\tilde{C}^4-86\tilde{S}^2\tilde{C}^2+60\tilde{S}^4\right)\right),\end{aligned} \tag{A.18}$$

$$\begin{aligned}\tau_{q7} \cong\ &16\tilde{S}\tilde{C}^2\left(1+\eta^2\right)\left(17\tilde{C}^4-26\tilde{S}^2\tilde{C}^2+2\tilde{S}^4+\eta^2\left(214\tilde{C}^4-562\tilde{S}^2\tilde{C}^2+124\tilde{S}^4\right)\right.\\ &\left.+\,\eta^4\left(815\tilde{C}^4-2980\tilde{S}^2\tilde{C}^2+1080\tilde{S}^4\right)\right).\end{aligned} \tag{A.19}$$

Appendix B: Formulae for Calculating the Coefficients $D_1, D_2, \ldots D_9, H_1$

$$D_1 = (d_1)_q - d_1\tau_q, \tag{B.1}$$

$$D_2 = (d_1)_\lambda + (d_2)_q - 2d_2\tau_q, \tag{B.2}$$

$$D_3 = (d_2)_\lambda + (d_3)_q + 3\tau_q(d_1 - d_3), \tag{B.3}$$

$$D_4 = (d_3)_\lambda + (d_4)_q + \tau_q(2d_2 - 4d_4), \tag{B.4}$$

$$D_5 = (d_4)_\lambda + d_3\tau_q, \tag{B.5}$$

$$H_1 = \left(h_q\tau_q + h\tau_{qq} - 2h\tau_q^2\right)\sin\alpha\cos\alpha + h_\lambda\tau_q\sin^2\alpha, \tag{B.6}$$

where

$$(d_1)_q = N_q E_{qqq} - E_q N_{qqq}, \tag{B.7}$$

$$(d_1)_\lambda + (d_2)_q = 3N_qE_{qq\lambda} - 3E_qN_{qq\lambda} + N_\lambda E_{qqq} - E_\lambda N_{qqq}, \tag{B.8}$$

$$(d_2)_\lambda + (d_3)_q = 3\big(N_qE_{q\lambda\lambda} - E_qN_{q\lambda\lambda} + N_\lambda E_{qq\lambda} - E_\lambda N_{qq\lambda}\big), \tag{B.9}$$

$$(d_3)_\lambda + (d_4)_q = N_qE_{\lambda\lambda\lambda} - E_qN_{\lambda\lambda\lambda} + 3N_\lambda E_{q\lambda\lambda} - 3E_\lambda N_{q\lambda\lambda}, \tag{B.10}$$

$$(d_4)_\lambda = N_\lambda E_{\lambda\lambda\lambda} - E_\lambda N_{\lambda\lambda\lambda}, \tag{B.11}$$

$$D_6 = e_q, \tag{B.12}$$

$$D_7 = 2f_q + e_\lambda - 2f\tau_q, \tag{B.13}$$

$$D_8 = g_q + 2f_\lambda + 2\tau_q(e - g), \tag{B.14}$$

$$D_9 = g_\lambda + 2f\tau_q. \tag{B.15}$$

The derivatives of the Jacobian and Gaussian coefficients included in the Formulae (B.6, B.12–B.15) are:

$$\begin{aligned} h_q &= N_{qq}E_\lambda + N_qE_{q\lambda} - N_{q\lambda}E_q - N_\lambda E_{qq}, \\ h_\lambda &= N_{q\lambda}E_\lambda + N_qE_{\lambda\lambda} - N_{\lambda\lambda}E_q - N_\lambda E_q, \end{aligned} \tag{B.16}$$

$$e_q = 2N_qN_{qq} + 2E_qE_{qq}, \quad e_\lambda = 2N_qN_{q\lambda} + 2E_qE_{q\lambda}, \tag{B.17}$$

$$\begin{aligned} f_q &= N_{qq}N_\lambda + N_qN_{q\lambda} + E_{qq}E_\lambda + E_qE_{q\lambda} \\ f_\lambda &= N_{q\lambda}N_\lambda + N_qN_{\lambda\lambda} + E_{q\lambda}E_\lambda + E_qE_{\lambda\lambda}, \end{aligned} \tag{B.18}$$

$$g_q = 2N_\lambda N_{q\lambda} + 2E_\lambda E_{q\lambda}, \quad g_\lambda = 2N_\lambda N_{\lambda\lambda} + 2E_\lambda E_{\lambda\lambda}. \tag{B.19}$$

Appendix C: The Best Projection from the Set of Close-to-Equal-Area Azimuthal Projections

Consider an azimuthal projection in which the function ρ is equal to:

$$\rho = 4R\cot\frac{z}{2}\ln\sec\frac{z}{2}. \tag{C.1}$$

Let us differentiate Formula (C.1), and we obtain:

$$\rho_z = -2R\csc^2\frac{z}{2}\ln\sec\frac{z}{2} + 2R. \quad \text{(C.2)}$$

Taking into account (5.1), we find the extremal linear scale factors of the projection.

$$\begin{aligned}\mu_1 &= 2 - 2\csc^2\tfrac{z}{2}\ln\sec\tfrac{z}{2}\\ \mu_2 &= 2\csc^2\tfrac{z}{2}\ln\sec\tfrac{z}{2}.\end{aligned} \quad \text{(C.3)}$$

Let us find the function $(a + b - 2)$:

$$a + b - 2 = \mu_1 + \mu_2 - 2 = 0. \quad \text{(C.4)}$$

Thus, the projection under consideration belongs to the set of projections that are close-to-equal-area.

According to Eqs. (5.34–5.37), the function β of all azimuthal projections is equal to the azimuth taken with the opposite sign. Using Eqs. (5.2–5.3), the azimuth can be considered as one of the isometric coordinates for azimuthal projections.In this case, the derivatives of β with respect to the isometric coordinates are determined by the system (5.38), which indicates that β is a harmonic function for all azimuthal projections. Taking into account equalities (4.13) and (5.32), we find its conjugate function $\ln r(1 + L)$:

$$\ln r(1 + L) = \ln\left(2k_1R\tan\frac{z}{2}\right), \quad \text{(C.5)}$$

where k_1 is a constant determined by Eq. (5.27).

Let us show that condition (4.16) is satisfied at the boundary. Using (C.5), we find the function L.

$$L = k_1\sec^2\frac{z}{2} - 1, \quad \text{(C.6)}$$

or at the boundary:

$$L = 2\csc^2\frac{z_0}{2}\ln\sec\frac{z_0}{2} - 1. \quad \text{(C.7)}$$

Considering Eq. (C.3), we obtain:

$$(a - b) = \mu_2 - \mu_1 = 4\csc^2\frac{z_0}{2}\ln\sec\frac{z_0}{2} - 2. \quad \text{(C.8)}$$

Comparing Eqs. (C7–C8), we conclude that the projection under consideration satisfies the boundary condition (4.16).

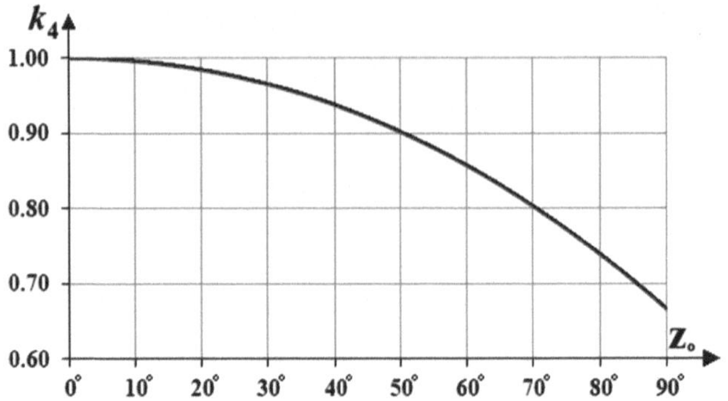

Fig. C.1 Graph of changes of parameter k_4 depending on z_0

Since Formulae (5.37 and 5.41) are valid for all azimuthal projections, taking into account all of the abovementioned projections, we can conclude that the projection under consideration is the best in the set of close-to-equal-area projections.

As is known (Bugayevskiy and Snyder 1995, p. 102), the polar distance of the best conformal azimuthal projection can be written as follows:

$$\rho = 2Rk_4 \tan\frac{z}{2}, \tag{C.9}$$

where k_4 is a certain constant determined by the condition that at the boundary, the linear distortion is equal to the value H, and at the central point of the projection, it is equal to the value $(-H)$. Using Eq. (5.1), we obtain for a conformal projection:

$$a = k_4 \sec^2\frac{z}{2}. \tag{C.10}$$

Hence:

$$k_4 = \frac{2\cos^2\frac{z_0}{2}}{1+\cos^2\frac{z_0}{2}}. \tag{C.11}$$

Figure C.1 shows a graph of k_4 as a function of z_0.

Let us take the arithmetic mean of expressions (C.1) and (C.9):

$$\rho = R\left[2\cot\frac{z}{2}\ln\sec\frac{z}{2} + k_4\tan\frac{z}{2}\right]. \tag{C.12}$$

According to Eqs. (C.12) and (5.27), the difference between k_1 and k_4 is:

$$k_5 = k_1 - k_4 = 2\cot^2\frac{z_0}{2}\ln\sec\frac{z_0}{2} - \frac{2\cos^2\frac{z_0}{2}}{1+\cos^2\frac{z_0}{2}}. \tag{C.13}$$

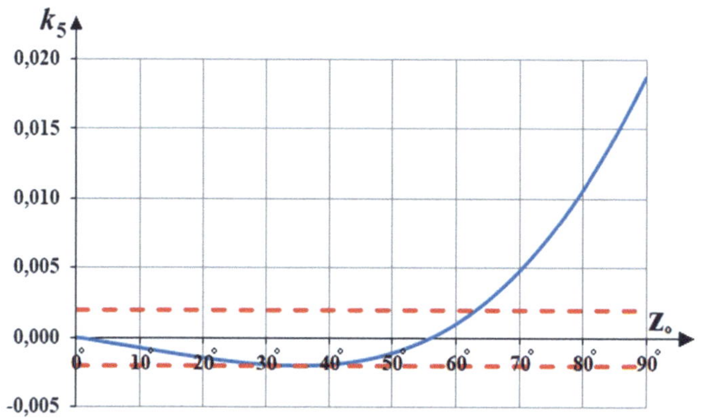

Fig. C.2 Graph of k_5 depending on z_0

Figure C.2 shows a graph of changes in the value of k_5 depending on z_0. According to this graph, we can conclude that if the linear distortions are small ($z_0 \leq 63$), then the difference between the values k_1 and k_4 does not exceed 0.002.

In this case, the Airy ideal projection can be obtained by finding the arithmetic mean of the mapping functions of the best conformal and the best of the set of close-to-equal-area projections. As mentioned above, the best equal-area projection and the best of the set of close-to-equal-area projection differ by quantities of the second order of smallness when the linear distortions are of the first order of smallness. Therefore, if the mapping functions of the best conformal and equal-area projections are known for any mapping area, then the mapping functions of the Airy ideal projection can be approximately obtained by taking the arithmetic mean of the mapping functions of the aforementioned best projections.

Applying this principle to azimuthal projections, we obtain the following approximate formula for the Airy ideal azimuthal projection:

$$\rho = R\left[\sin\frac{z}{2} + k_4 \tan\frac{z}{2}\right]. \qquad \text{(C.14)}$$

The boundary almucantar with a zenith distance z_0 for the best conformal projection is represented by a circle of radius:

$$\rho_0 = 2Rk_4 \tan\frac{z_0}{2}. \qquad \text{(C.15)}$$

The boundary almucantar with zenith distance z_0 for the best projection of the set of close-to-equal-area projections is represented by the circle of radius:

$$\rho_0' = 4R\cot\frac{z_0}{2}\ln\sec\frac{z_0}{2} = 2Rk_1 \tan\frac{z_0}{2}. \qquad \text{(C.16)}$$

The boundary almucantar for the Airy projection is equal to:

Table D.1 Parameters of the WGS84 Ellipsoid (NGA.STND.0036_1.0.0_WGS84 2014)

Parameter	Notation	Value
Semi-major axis (meters)	a	6,378,137.0
Reciprocal of flattening	$1/f$	298.257223563
First eccentricity	e	8.1819190842622E−02
First eccentricity squared	e^2	6.69437999014E−03
Second eccentricity	e'	8.2094437949696E−02
Second eccentricity squared	e'^2	6.73949674227510E−03

$$\rho_0'' = R\left[2\cot\frac{z_0}{2}\ln\sec\frac{z_0}{2} + k_1\tan\frac{z_0}{2}\right] = 2Rk_1\tan\frac{z_0}{2}. \tag{C.17}$$

If the linear distortions are quantities of the first order of smallness, then, as mentioned above:

$$k_1 \approx k_4. \tag{C.18}$$

Therefore, taking into account Formulae (C.15–C.18), we can write:

$$\rho_0'' = \rho_0' \approx \rho_0. \tag{C.19}$$

This finding indicates that the images of the boundary of the mapping area in the three projections under consideration coincide.

Appendix D: Geometric Parameters of the WGS84 Ellipsoid

See Table D.1.

Appendix E: Working Formulae and Parameters for Calculating the Meridian Arc Length and the Area of an Ellipsoidal Trapezoid

The length of the meridian arc is determined by the following formula:

$$S = \int_0^{q_0} r\mathrm{d}q = \int_0^{\varphi_0} M\,\mathrm{d}\varphi. \tag{E.1}$$

According to (Rapp 1991, p. 38), to calculate the length of the meridian arc from the equator to the point of latitude φ_0, the following formula is used:

$$S = a\left(1-e^2\right)\left(A\varphi_0 - \frac{B}{2}\sin 2\varphi_0 + \frac{C}{4}\sin 4\varphi_0 - \frac{D}{6}\sin 6\varphi_0 + \frac{E}{8}\sin 8\varphi_0 - \right) + \cdots, \tag{E.2}$$

where $A, B, \ldots F$ are the parameters associated with the first eccentricity according to the following formulae:

$$\begin{aligned} A &= 1 + \tfrac{3}{4}e^2 + \tfrac{45}{64}e^4 + \tfrac{175}{256}e^6 + \tfrac{11025}{16384}e^8 + \tfrac{43659}{65536}e^{10} \\ B &= \tfrac{3}{4}e^2 + \tfrac{15}{16}e^4 + \tfrac{525}{512}e^6 + \tfrac{2205}{2048}e^8 + \tfrac{72765}{65536}e^{10} \\ C &= \tfrac{15}{64}e^4 + \tfrac{105}{256}e^6 + \tfrac{2205}{4096}e^8 + \tfrac{10395}{16384}e^{10} \\ D &= \tfrac{35}{512}e^6 + \tfrac{315}{2048}e^8 + \tfrac{31185}{131072}e^{10} \\ E &= \tfrac{315}{16384}e^8 + \tfrac{3465}{65536}e^{10} \\ F &= \tfrac{693}{131072}e^{10} \end{aligned} \quad . \tag{E.3}$$

The area of a spheroidal trapezoid between the equator and the parallel with latitude φ_0 and between two meridians with a longitude difference of 1 rad is determined by the following equation:

$$P = \int_0^{q_0} r^2 \mathrm{d}q = \int_0^{\varphi_0} Mr\mathrm{d}\varphi. \tag{E.4}$$

or, according to (Rapp 1991, p. 43):

$$P = a^2\left(1-e^2\right)\left(\sin\varphi_0 + \frac{2}{3}e^2\sin^3\varphi_0 + \frac{3}{5}e^4\sin^5\varphi_0 + \frac{4}{7}e^6\sin^7\varphi_0 + \frac{5}{9}e^8\sin^9\varphi_0 + \cdots\right). \tag{E.5}$$

Appendix F: Mapping Functions and Their Derivatives for the Ideal According to Airy Transverse Cylindrical Projection

F.1 Difference Between the Mapping Function N' of an Ideal According to Airy Transverse Cylindrical Projection and the Corresponding Gauss–Kruger Projection

For the functions included in Formula (7.54), we can write:

$$\Delta N_{G-K} = \overline{N}_4 l^4 + \overline{N}_6 l^6 + \overline{N}_8 l^8, \tag{F.1}$$

where

$$\overline{N}_4 = \frac{1}{6} r\eta^2 \sin\varphi \cos^2\varphi (1 + \eta^2), \tag{F.2}$$

$$\overline{N}_6 = \frac{1}{180} r\eta^2 \tilde{S}\tilde{C}^2 (31\tilde{C}^2 - 35\tilde{S}^2 - \eta^2(\tilde{C}^2 + 23\tilde{S}^2) + \eta^4(19\tilde{C}^2 - 6\tilde{S}^2) + 10\eta^6\tilde{C}^2), \tag{F.3}$$

$$\begin{aligned} \overline{N}_8 = \frac{1}{10080} r\eta^2 \tilde{S}\tilde{C}^2 & (2629\tilde{C}^4 - 1183\tilde{S}^2\tilde{C}^2 + 1760.5\tilde{S}^4 \\ & + \eta^2(3304\tilde{C}^4 - 2895\tilde{S}^2\tilde{C}^2 + 1687\tilde{S}^4) \\ & + \eta^4(1977\tilde{C}^4 - 7408\tilde{S}^2\tilde{C}^2 + 3054\tilde{S}^4) \\ & + \eta^6(1873\tilde{C}^4 - 7694\tilde{S}^2\tilde{C}^2 + 2064\tilde{S}^4)). \end{aligned} \tag{F.4}$$

F.2 *Functions of Latitude, Included in the Formula for the Mapping Function E′ of the Ideal According to Airy Transverse Cylindrical Projection*

The mapping function E', determined from the second equation of the system (7.4), contains the following functions depending on latitude:

$$E'_1 = r, \tag{F.5}$$

$$E'_3 = -\frac{1}{6} r \sin^2\varphi, \tag{F.6}$$

$$E'_5 = \frac{1}{120} r(\tilde{S}^4 - 8\tilde{S}^2\tilde{C}^2 - 4\eta^2\tilde{C}^2 + 4\eta^4\tilde{C}^2(7\tilde{S}^2 - 2\tilde{C}^2) + 4\eta^6\tilde{C}^2(6\tilde{S}^2 - \tilde{C}^2)), \tag{F.7}$$

$$\begin{aligned} E'_7 \cong \frac{1}{5040} r & (-\tilde{S}^6 + 88\tilde{S}^4\tilde{C}^2 - 136\tilde{S}^2\tilde{C}^4 \\ & + \eta^2\tilde{C}^2(-124\tilde{C}^4 + 140\tilde{S}^4 + 328\tilde{S}^2\tilde{C}^2) \\ & + \eta^4\tilde{C}^2(-2420\tilde{S}^4 + 4640\tilde{S}^2\tilde{C}^2 - 548\tilde{C}^4) \\ & + \eta^6\tilde{C}^2(11016\tilde{S}^2\tilde{C}^2 - 8616\tilde{S}^4 - 900\tilde{C}^4)). \end{aligned} \tag{F.8}$$

F.3 Derivatives with Respect to q of the Mapping Functions

The derivatives of the functions N_4', N_6', N_8' have the following form:

$$N'_{4q} = \frac{1}{6} r\eta^2 \tilde{C}^2 \left(\tilde{C}^2 - 5\tilde{S}^2 + \eta^2 \left(2\tilde{C}^2 - 5\tilde{S}^2\right) + \eta^4 \tilde{C}^2\right), \tag{F.9}$$

$$\begin{aligned} N'_{6q} \cong \frac{1}{180} r\eta^2 \tilde{C}^2 &\left(31\tilde{C}^4 - 322\tilde{S}^2\tilde{C}^2 + 175\tilde{S}^4\right. \\ &- \eta^2\left(30\tilde{C}^4 - 351\tilde{S}^2\tilde{C}^2 + 301\tilde{S}^4\right) \\ &\left. + \eta^4\left(18\tilde{C}^4 - 288\tilde{S}^2\tilde{C}^2 + 192\tilde{S}^4\right)\right), \end{aligned} \tag{F.10}$$

$$\begin{aligned} N'_{8q} \cong \frac{1}{10080} r\eta^2 \tilde{C}^2 &\left(2629\tilde{C}^6 - 27210\tilde{S}^2\tilde{C}^4 + 17083.5\tilde{S}^4\tilde{C}^2 - 8802.5\tilde{S}^6\right. \\ &+ \eta^2\left(5933\tilde{C}^6 - 69610\tilde{S}^2\tilde{C}^4 + 50390.5\tilde{S}^4\tilde{C}^2 - 18851\tilde{S}^6\right) \\ &\left. + \eta^4\left(5281\tilde{C}^6 - 89650\tilde{S}^2\tilde{C}^4 + 128353\tilde{S}^4\tilde{C}^2 - 37608\tilde{S}^6\right)\right). \end{aligned} \tag{F.11}$$

Similarly, with respect to q, the derivatives of the functions E_1', E_3', E_5' are equal to:

$$E'_{1q} = -r \sin\varphi, \tag{F.12}$$

$$E'_{3q} = -\frac{1}{3!} r \sin\varphi \left(-2\cos^2\varphi + \sin^2\varphi - 2\eta^2 \cos^2\varphi\right), \tag{F.13}$$

$$\begin{aligned} E'_{5q} \cong \frac{1}{5!} r\tilde{S} &\left(-\tilde{S}^4 + 28\tilde{S}^2\tilde{C}^2 - 16\tilde{C}^4 + 4\eta^2\tilde{C}^2\left(10\tilde{S}^2 + \tilde{C}^2\right)\right. \\ &\left. + 36\eta^4\tilde{C}^2\left(-5\tilde{S}^2 + 4\tilde{C}^2\right) + 4\eta^6\tilde{C}^2\left(53\tilde{C}^2 - 96\tilde{S}^2\right)\right), \end{aligned} \tag{F.14}$$

$$\begin{aligned} E'_{7q} \cong \frac{1}{7!} r\tilde{S} &\left(\tilde{S}^6 - 270\tilde{S}^4\tilde{C}^2 - 1032\tilde{S}^2\tilde{C}^4 - 272\tilde{C}^6\right. \\ &+ \eta^2\tilde{C}^2\left(-882\tilde{S}^4 - 840\tilde{S}^2\tilde{C}^2 + 1500\tilde{C}^4\right) \\ &+ \eta^4\tilde{C}^2\left(-52848\tilde{S}^2\tilde{C}^2 + 16956\tilde{C}^4 + 16380\tilde{S}^4\right) \\ &\left. + \eta^6\tilde{C}^2\left(48492\tilde{C}^4 - 202440\tilde{S}^2\tilde{C}^2 + 92064\tilde{S}^4\right)\right). \end{aligned} \tag{F.15}$$

Appendix G: Mapping Functions and Their Derivatives with Respect to q for the Cassini–Soldner Projection

The mapping functions of the Cassini–Soldner projection are as follows:

$$N_{C-S} = N_0'' + N_2''l^2 + N_4''l^4, \tag{G.1}$$

$$E_{C-S} = E_1''l + E_3''l^3 + E_5''l^5, \tag{G.2}$$

where

$$\begin{aligned} N_0'' &= S \\ N_2'' &= \tfrac{1}{2!} r \sin\varphi \\ N_4'' &= \tfrac{1}{4!} r \sin\varphi \cos^2\varphi\left(5 - \tan^2\varphi + 6e'^2\cos^2\varphi\right), \end{aligned} \tag{G.3}$$

$$\begin{aligned} E_1'' &= r \\ E_3'' &= -\tfrac{1}{3!} r \sin^2\varphi \\ E_5'' &= -\tfrac{1}{5!} r \sin^2\varphi\cos^2\varphi\left(8 - \tan^2\varphi + 8e'^2\cos^2\varphi\right). \end{aligned} \tag{G.4}$$

The derivatives with respect to q of the functions N_0'', N_2'', N_4'' have the following form:

$$N_{0q}'' = r, \quad N_{2q}'' = -\frac{1}{2!} r\left(\sin^2\varphi - \cos^2\varphi -^2 \cos^2\varphi\right),$$

$$N_{4q}'' = \frac{1}{4!} r\left(5\tilde{C}^4 - 18\tilde{S}^2\tilde{C}^2 + \tilde{S}^4 + \eta^2\tilde{C}^2\left(11\tilde{C}^2 - 43\tilde{S}^2\right) + \eta^4\tilde{C}^2\left(6\tilde{C}^2 - 24\tilde{S}^2\right)\right), \tag{G.5}$$

Similarly, the derivatives with respect to q of the functions E_1'', E_3'', E_5'' are equal to:

$$E_{1q}'' = -r\sin\varphi, \quad E_{3q}'' = -\frac{1}{3!} r\sin\varphi\left(-2\cos^2\varphi + \sin^2\varphi - 2\eta^2\cos^2\varphi\right),$$

$$E_{5q}'' = -\frac{1}{5!} r\tilde{S}\left(16\tilde{C}^4 - 28\tilde{S}^2\tilde{C}^2 + \tilde{S}^4 + \eta^2\tilde{C}^2\left(32\tilde{C}^2 - 60\tilde{S}^2\right) + 16\eta^4\tilde{C}^2\left(\tilde{C}^2 - 2\tilde{S}^2\right)\right). \tag{G.6}$$

Appendix H: Working Formulae for Constructing the Markov Projection

1. The formulae for determining the latitudes φ_3, φ_4 are as follows:

$$\frac{1}{M_3} r_\varphi(\varphi_3) \int_{\varphi_1}^{\varphi_3} M \, d\varphi + \frac{M_3}{M_4} \cdot \frac{r_\varphi(\varphi_4)}{r_\varphi(\varphi_3)} r_1 - r_3 = 0, \tag{H.1}$$

$$-\frac{1}{M_3} r_\varphi(\varphi_3) \int_{\varphi_4}^{\varphi_2} M \, d\varphi + \frac{M_4}{M_3} \cdot \frac{r_\varphi(\varphi_3)}{r_\varphi(\varphi_4)} r_2 - r_4 = 0. \tag{H.2}$$

2. General parameters:

$$\kappa = \sin \varphi_3, \tag{H.3}$$

$$(1 + H)^2 = \frac{\sin \varphi_4}{\sin \varphi_3}. \tag{H.4}$$

3. Formulae for calculating the radii of parallels on a plane and the scale factors of Zone I are as follows:

$$\dot{\rho} = -\frac{1}{1+H} \int_{\varphi_1}^{\varphi} M \, d\varphi + \frac{(1+H)}{\alpha} r(\varphi_1), \tag{H.5}$$

$$\dot{\rho} = \frac{1}{1+H} \int_{\varphi}^{\varphi_3} M \, d\varphi + \frac{1}{\alpha(1+H)} r(\varphi_3), \tag{H.6}$$

$$m = \frac{1}{1+H} = \text{const}, \tag{H.7}$$

$$n = \frac{\dot{\rho}}{r}. \tag{H.8}$$

4. The formulae for calculating the radii of parallels on a plane and the scale factors of Zone II are as follows:

$$\ddot{\rho} = \frac{1}{(1+H)} r, \tag{H.9}$$

$$\ddot{\rho}_\varphi = \frac{1}{(1+H)} r_\varphi, \tag{H.10}$$

$$m = -\frac{\ddot{\rho}_\varphi}{M}, \tag{H.11}$$

$$n = \frac{\ddot{\rho}}{r} = \frac{1}{1+H} = \text{const}. \tag{H.12}$$

5. Formulae for calculating the radii of parallels on a plane and the scale factors of Zone III are as follows:

$$\dddot{\rho} = (1 + H) \int\limits_{\varphi}^{\varphi_2} M \, \mathrm{d}\varphi + \frac{1}{\kappa}(1 + H) r(\varphi_2), \tag{H.13}$$

$$\dddot{\rho} = -(1 + H) \int\limits_{\varphi_4}^{\varphi} M \, \mathrm{d}\varphi + \frac{1}{\kappa(1 + H)} r(\varphi_4), \tag{H.14}$$

$$m = -\frac{\dddot{\rho}_{\varphi}}{M} = (1 + H) = \text{const}, \tag{H.15}$$

$$n = \kappa \frac{\dddot{\rho}}{r}. \tag{H.16}$$

Appendix I: Working Formulae for the Lambert, Albers, and Equidistant Conic Projections, Determined by the Chebyshev–Markov Criterion

The projection parameters are determined from the following conditions:

1. At the boundary parallels with latitudes φ_1, φ_2, the maximum scale factors are equal to $(1 + H)$, namely:

$$a_1 = a_2 = 1 + H. \tag{I.1}$$

2. At the central parallel of the projection with latitude φ_0, the minimum scale factor is equal to:

$$b_0 = \frac{1}{1 + H}. \tag{I.2}$$

Since the minimum of the linear scale factor is reached at this parallel, the derivative of this quantity is zero:

$$\left(\frac{db}{d\varphi}\right)_0 = 0. \tag{I.3}$$

According to (Bugayevskiy and Snyder 1995, pp. 90, 93, 95), the result of solving Eq. (I.3) is written as follows:

$$\overline{\kappa} = \sin \varphi_0. \tag{I.4}$$

All the abovementioned projections have 4 parameters: $\overline{\kappa}$, $\overline{k}$, H, φ_0. To find them, there are 4 conditions (I.1–I.4). Let us consider the determination of these parameters for specific projections.

I.1 Lambert Projection

The linear scale factors of this projection are equal (Bugayevskiy and Snyder 1995, p. 90):

$$m = a = b = \frac{\overline{\kappa}\overline{k}}{rU^{\overline{\kappa}}}, \tag{I.5}$$

where U is a function of latitude, determined by Formula (1.10). Substituting Eq. (I.5) into the first formula of equalities (I.1) yields:

$$\overline{\kappa} = \frac{\ln r_1 - \ln r_2}{\ln U_2 - \ln U_1}. \tag{I.6}$$

Multiplying Eq. (I.1) with Eq. (I.2) and taking the square root of the resulting expression, we have:

$$\overline{k} = \frac{\sqrt{r_0 r_1 U_0^{\overline{\kappa}} U_1^{\overline{\kappa}}}}{\overline{\kappa}} = \frac{\sqrt{r_0 r_2 U_0^{\overline{\kappa}} U_2^{\overline{\kappa}}}}{\overline{\kappa}}, \tag{I.7}$$

where r_0 is the radius of the parallel with latitude φ_0; U_0 is a function of the same latitude, which is calculated using Formula (1.10). The following equations can be used to determine parameter H:

$$H = \frac{\overline{k}}{r_1 U_1} - 1 = \frac{\overline{k}}{r_2 U_2} - 1 = \frac{r_0 U_0}{\overline{k}} - 1. \tag{I.8}$$

I.2 Albers Projection

The linear scale factors of the Albers projection are equal (Bugayevskiy & Snyder, 1995, p. 93):

$$n = \frac{1}{m} = \frac{\sqrt{2\left(\overline{k} - P\right)}}{r}, \tag{I.9}$$

where P is the area of a spheroidal trapezoid between the equator and a parallel with a certain latitude and two meridians with a longitude difference of one radian. The value of P can be calculated using Eq. (E.5) in Appendix E.

For the conditions associated with the Chebyshev–Markov criterion, the following equalities can be written:

$$\frac{\sqrt{2\left(\bar{k}-P_1\right)}}{r_1}=\frac{\sqrt{2\left(\bar{k}-P_2\right)}}{r_2}=1+H, \tag{I.10}$$

$$\frac{\sqrt{2\left(\bar{k}-P_0\right)}}{r_0}=\frac{1}{1+H}. \tag{I.11}$$

Using the first part of equalities (I.10), we obtain the following formula for calculating the parameter $\bar{k}$.

$$\bar{k}=\frac{P_2 r_1^2-P_1 r_2^2}{r_1^2-r_2^2}. \tag{I.12}$$

Multiplying Eq. (I.11) by the first and then by the second formula in equalities (I.10), we obtain:

$$\frac{\sqrt{\bar{k}-S_0}}{r_0}=\frac{r_1}{2\sqrt{\bar{k}-S_1}}=\frac{r_2}{2\sqrt{\bar{k}-S_2}}. \tag{I.13}$$

The functions on the left side of equation (I.13) are dependent only on latitude φ_0. Both equations (I.13) are transcendental. Therefore, the latitude φ_0 can be found using one of the well-known approximate methods, for example, Newton's method (Atkinson, 1989).

To determine the parameter H, the following equalities are used:

$$H=\frac{\sqrt{2\left(\bar{k}-P_1\right)}}{r_1}-1=\frac{\sqrt{2\left(\bar{k}-P_2\right)}}{r_2}-1=\frac{r_0}{\sqrt{2\left(\bar{k}-P_0\right)}}-1. \tag{I.14}$$

I.3 Conical Projection Equidistant Along the Meridians

The linear scale factor along the parallels of the projection equidistant along the meridians is given by (Bugayevskiy & Snyder, 1995, p. 95):

$$n = \frac{\left(\overline{k} - S\right)}{r}, \tag{I.15}$$

where S is the length of the meridian arc from the equator to the parallel at a certain latitude, which can be calculated using Eq. (E.2) from Appendix E.

The boundary conditions of this projection are as follows:

$$\frac{\left(\overline{k} - S_1\right)}{r_1} = \frac{\left(\overline{k} - S_2\right)}{r_2} = 1 + H. \tag{I.16}$$

Equality (I.2) for this projection is as follows:

$$\frac{\left(\overline{k} - S_0\right)}{r_0} = \frac{1}{1 + H}. \tag{I.17}$$

The parameter $\overline{k}$, taking into account equalities (I.16), is equal to:

$$\overline{k} = \frac{S_2 r_1 - S_1 r_2}{r_1 - r_2}. \tag{I.18}$$

To determine the parameter φ_0, transcendental equations similar to equations (I.13) are used:

$$\frac{\overline{k}^2\left(\overline{k} - S_0\right)}{r_0} = \frac{r_1}{\overline{k} - S_1} = \frac{r_2}{\overline{k} - S_2}, \tag{I.19}$$

which are also solved via approximate methods, such as Newton's method (Atkinson 1989).

The value of the Chebyshev–Markov criterion is determined as follows:

$$H = \frac{\left(\overline{k} - S_1\right)}{r_1} - 1 = \frac{\left(\overline{k} - S_2\right)}{r_2} - 1 = \frac{r_0}{\left(\overline{k} - S_0\right)} - 1. \tag{I.20}$$

References

Atkinson KE (1989) An introduction to numerical analysis, 2nd edn. Wiley, p 663. ISBN 0-471-50023-2. https://www.math.science.cmu.ac.th/docs/qNA2556/ref_na/Katkinson.pdf

Bagratuni GV (1962) Course in spheroidal geodesy, Technical Report FTD-MT-64-390, US Air Force, translation of Kurs sferoidicheskoi geodezii, 1967. (Geodezizdat, Moscow, 1962), by Foreign Technology Division, Wright-Patterson AFB, p 305. https://apps.dtic.mil/sti/pdfs/AD0650520.pdf

Bugayevskiy LM, Snyder JP (1995) Map projections. Reference manual. Taylor&Francis Ltd., London, p 328. https://doi.org/10.1201/b16431

NGA.STND.0036_1.0.0_WGS84 (2014) National Geospatial-Intelligence Agency (NGA) Standardization Document Department of Defense World Geodetic System 1984. 2014-07-08 Version 1.0.0. Office of Geomatics. https://nsgreg.nga.mil/doc/view?i=4085

Rapp R (1991) Geometric geodesy. Part I. The Ohio State University. Department of Geodetic Science and Surveying, Columbus, Ohio, p 189. https://geodesyattamucc.pbworks.com/f/Rapp_Geom_Geod_Vol_I.pdf

E. Novikova, *Best Map Projections*, https://doi.org/10.1007/978-3-031-78334-0

Index

E. Novikova, *Best Map Projections*, https://doi.org/10.1007/978-3-031-78334-0

The manufacturer's authorised representative in the EU is Springer
Nature Customer Service Centre GmbH, Europaplatz 3, 69115 Heidelberg,
Germany. If you have any concerns regarding our products, please
contact ProductSafety@springernature.com

Printed and bound by CPI Group (UK) Ltd, Croydon, CR0 4YY
30/09/2026
02225568-0003